# Strategies for Technical Writing

*A Rhetoric with Readings*

# Strategies for Technical Writing

## A Rhetoric with Readings

**Mary M. Lay**
*Clarkson College*

**Holt, Rinehart and Winston**
New York   Chicago   San Francisco   Philadelphia
Montreal   Toronto   London   Sydney
Tokyo   Mexico City   Rio de Janeiro   Madrid

**Library of Congress Cataloging in Publication Data**
Lay, Mary M.
  Strategies for technical writing.
  Includes index.
  1. English language—Rhetoric. 2. College readers.
3. English language—Business English. I. Title.
PR1408.L348      808′.0666021      81-6802
                                           AACR2
**ISBN 0-03-053636-7**

Copyright © 1982 by CBS College Publishing
Address correspondence to:
383 Madison Avenue
New York, N.Y. 10017
All rights reserved
Printed in the United States of America
Published simultaneously in Canada
2  3  4  5  059  9 8 7 6 5 4 3 2
CBS COLLEGE PUBLISHING
Holt, Rinehart and Winston
The Dryden Press
Saunders College Publishing

# Credits

p. 10, extract, Max Gunther, "How Television Helps Johnny Read," *TV Guide*, September 4, 1976, p. 9.

pp. 10–11, extract, Bob Tripolsky, "We Test the 1977 Plymouth," *Mechanics Illustrated*, September 1976, p. 48.

p. 11, extract, Myron B. Bloy, Jr., "Hardness of Soul, Sclerosis of Imagination," *The Chronicle of Higher Education*, 1976, back page.

p. 12, extract, "Cooling Coils," "Basic Engineering Report," Syska and Hennessy, Inc., Engineers. An example of upward communication.

pp. 12–13, extract, Kenneth F. Neusen, "High Velocity Fluid Jet Cutting and Slotting," Technical Paper, Society of Manufacturing Engineers, 1976. An example of lateral communication.

p. 13, extract, "Communications" Guide, courtesy of Corning Glass Works, Technical Information Center. An example of downward communication.

p. 14, extract, *General Catalog 501*, Fisher Controls, Co. Reprinted with the permission of Fisher Controls Co. Copyright © by Fisher Controls, 1975. All rights reserved. An example of outward communication.

p. 16, Reading IA, "Cosmos 954: An Ugly Death," *Time*, Vol. 111, No. 6 (February 6, 1978), p. 28.

p. 16, Reading IB, "The Unscheduled Return of Cosmos 954," *Science News*, Vol. 113, No. 6 (February 4, 1978), p. 69. Reprinted with permission from *Science News*, the weekly news magazine of science, copyright 1978 by Science Service, Inc.

p. 17, Reading IC, Geoffrey Norman, "The Satellite That Invaded a Campsite," *Esquire*, Vol. 89, No. 4 (March 14, 1978), p. 85.

*(Credits are continued on page 299.)*

*For my parents,*
*Charles and Doris Lay*

# Preface

Today everyone involved in technology, science, and business must be able to organize and present information to a variety of readers or audiences for a variety of purposes. This book teaches the student just beginning a study of professional writing how to organize material according to basic rhetorical patterns common to professional writing. It also teaches students how to read more critically by judging why and how well other writers have organized their communications. Thus the book meets the needs of three types of courses: the Freshman English course in a school whose students major primarily in engineering, science, or business; the Applied Composition course that concentrates on professional communication situations; and the Technical Writing course that stresses audience analysis and organization before exploring report formats.

The book begins with an introduction to audience and purpose analysis—a technique stressed throughout the text. After an introduction to the special methods of gathering and organizing technical information, the book teaches the student how first to identify and then present a subject to an audience by definition and description. Then the student examines the organizational patterns most useful in presenting or analyzing more than one subject: comparison-contrast, causal analysis, and classification-partition. Once the student has learned how to organize basic subject matter, he or she then learns how to describe subjects in motion or use by narrative, process description, instructions, procedures, and specifications. Finally students explore how to convince an audience by argument and persuasion. The last two chapters introduce the formats common in professional writing—letters, memos, and reports—and how the organizational patterns described contribute to the design of those formats.

The book makes a special contribution to the growing field of technical communication for several reasons. It ensures that students begin technical writing courses with a thorough background in how to organize any communication according to its unique audience and purpose; also it ensures that students can apply what they traditionally learn in Freshman Composition. Thus they learn the practical use of rhetorical concepts in professional situations. Most chapters begin with a "classical" example of the organizational pattern so that it can be seen how useful a technique has been for perhaps hundreds of years. Then the special characteristics of an organizational pattern are examined, such as the differences between a formal and a stipulatory definition or the dangers of identifying only one cause of a phenomenon. The applications of each organizational pattern appear next, followed by any special stylistic characteristics or graphic aids unique to the pattern. Each section of each chapter is illustrated by excerpts from professional on-the-job writing, contemporary scientific, technical, or business journalism, and successful student papers. Usually two short and one long reading appear at the end of each chapter with key questions to help evaluate how well another writer organized his or her communication. Each chapter concludes with exercises and assignments designed to give students with various backgrounds and interests practice in writing.

This book was originally conceived to meet the specific needs of a course at Clarkson College: Theory of Rhetoric for Business, Science, and Engineering. Offered in the second semester of the Freshman year to students majoring in management, industrial distribution, biology, physics, chemistry, mathematics and computer science, chemical engineering, civil and environmental engineering, electrical and computer engineering, mechanical and industrial engineering, and technical communications, the course introduces students to both theory and practice in professional writing. Because no other book on the market offered substantial readings while teaching these students how to analyze and solve professional communication problems, I designed this one. Much of its content has been tested for the last three years in my and my colleagues' classrooms. We find that this approach to audience analysis and organizational patterns creates not only better writers but better thinkers and ensures that all students can understand and apply the traditional patterns and formats of professional writing.

## Acknowledgments

I would like to thank the following reviewers who made helpful criticisms and suggestions: Rosemary Ascherl, Hartford State Technical College; James Corey, New Mexico Tech; Albert J. Geritz, Fort Hays State University; John S. Harris, Brigham Young University; Audrey Hodgins, National Council of Teachers of English; Wayne A. Losano, University of Florida; Frances Maguire, Tarrant County Junior College; Kathleen O'Shaugnessy, Coastal Carolina Community College; Nell Ann Pickett, Hinds Junior College; Charles R. Stratton, University of Idaho; and Eugene Street, Brevard Community College.

Many thanks are due to my Rhetoric and Report Writing students for allowing me to use their work in preparing this text, and to my colleagues for helping me experiment with various approaches in the Rhetoric class and for saving sample papers for me. My appreciation is extended also to Bradford Broughton for his encouragement, to David Craig for his ideas, wisdom, and patience, and to Faye Serio who took the photographs, helped me with the artwork, and translated theory into art. Carolyn and Grant Gould created an atmosphere in which I could work quietly and productively, and Nancy Schermerhorn and Dorothy Mein typed accurately and quickly the early drafts of the manuscript. Mary Veglahn not only typed the final manuscript but often served as editor as well. My editor at Holt, Rinehart and Winston, Anne Boynton-Trigg, helped me maintain my vision throughout the long process of writing the book, and Peggy Middendorf saw the book through the production stage. Finally I would like to thank the following companies and institutions for providing me with samples of professional writing:

Applied Data Research, Inc.
Beckman Instruments, Inc.
California Department of Transportation
Campbell Engineering Company
Cincom Systems, Inc.
Corning Glass Works
Dames & Moore
T.E. Davidson & Associates, Inc.
DeWild Grant Reckert & Associates
Eastman Kodak Company
Environmental Research Institute of Michigan
Firman Technical Publications, Inc.
Fisher Controls Company
S.R. Harper, Inc.
Harris Corporation
Hartman Sales Corp.
IBM Corporation

Institute for Scientific Information
International Minerals & Chemical Corporation
Lester B. Knight & Associates
Los Alamos Scientific Laboratory
New York State Health Research Council
B. Nolte & Sons, Inc.
Phillips Petroleum Company
Polysar Limited
Public Works Department, Kansas City, Missouri
Richardson, Runden & Company, Inc.
Stromberg-Carlson Corporation
Syska & Hennessy, Inc., Engineers
Westinghouse Electric Corporation
Xerox Corporation

M. L.
Potsdam, New York

# Contents

# PART THREE    Presenting Multiple Subjects to an Audience    81

# PART FOUR   Presenting Subjects in Motion or in Use to an Audience   141

**10  Instructions:** Directing How to Use or Repair a Device    181

**11  Procedures:** Standardizing Operations; **And Specifications:** Prescribing Criteria    198

# PART ONE

## Basic Concerns of Technical Communication

# Audience and Purpose Analysis

*Analyzing the audience, purpose, and direction*
*of a communication.*
*Choosing a pattern of organization.*
*Addressing an audience in an appropriate style.*
*Addressing the needs and interests of special audiences.*
*Analyzing the audience and purpose of communications*
*written by others.*

In technical communication, as in all communication, we present facts and ideas and persuade our audience of readers or listeners to accept or act upon these facts and ideas. However, in technical communication, we address a particular audience, often already interested in or knowledgeable about our subject. We address special audiences of supervisors, managers, technologists, public officials and interest groups, stockholders, potential customers, and research scientists and engineers. We tailor our communications to meet the needs and interests of our particular audience.

Because the content of technical communications comes from technology, science, and industry, it is specialized and often complex. To present this specialized subject matter clearly, concisely, and objectively, we use certain organizational or rhetorical patterns. These patterns help us define terms, describe devices at rest or in motion, contrast alternatives, analyze causes, classify items, give instructions, and meet the other requirements of technical communication.

Thus in technical communication, we address a particular audience with a special interest or knowledge. We present and recommend facts and ideas in a clear, concise, and logical way.

In this book, we study the organizational patterns we use in technical communication. We examine first the patterns we use to identify and describe our subject, then the patterns we use to present more than one subject and to describe our subjects in use or motion, and finally the patterns we use to convince an audience.

To understand each pattern, we examine its characteristics and then see how it can be used in communications for various audiences. Our examples come from classical scientific and technical literature, professional communications in science and industry, contemporary scientific and technical journalism, and communications written by technical students. Once we learn organizational patterns, we can design a technical communication to address the needs and interests of any audience.

## The Audience of a Communication

In technical communication, we analyze our audience to determine such things as what to say first, how much to say, what words to use, what organizational pattern to use, and even how to title a communication. Before we write the first word in the communication, we gather and analyze facts about the members of our audience. We do this not to manipulate or fool them, not to tell them just what they want to hear, but to present our message in the clearest, most concise and useful way. Why not just ask the members of the audience what they want and then give it to them? They may not know everything the writer has to say. They may not have the technical background the writer has. They may be busy. Very often the audience consists of decision-makers who determine what future company projects will be and who will run these projects; they will make a decision based on a number of communications from a number of writers. When we analyze our audience, we go beyond conveying statistics and results; we go beyond plugging information into a set format; we become decision-makers ourselves.

## Questions to Ask about Audiences

When analyzing our audience, we can ask such questions as the following:

1. What are the general characteristics of our audience? Age? Education? Professional background and experience? Position in the company? Primary responsibilities?
2. What do the members of the audience know about this specific subject? What do they know already and what is new to them?
3. What is the audience's value system? With what are its members most or least concerned?
4. What are the audience's reading habits or patterns? Do its members read all or parts of the communication? What would they want to know first? Last?
5. What might the audience use the communication for now and in the future?
6. Who else might read the communication? Are there more audiences than the one we sent the communication to ourselves?
7. Who might be affected by our communication? Whose job might be changed by our recommendations? How would this audience react to our suggestions?

The answers to these questions help us shape our communication so that the members of the audience understand it, value it, and can use or apply it to their jobs.

    The age of the audience determines such things as the examples we use; will the members of the audience remember an event such as the bombing of Hiroshima or will they have only read about it in history texts? The education and professional experience of the audience determine among other things the vocabulary we use; should we say "five times the explosive force dropped on Hiroshima" or should we say instead "100 pounds of uranium 235, a power output in the range of tens of hundreds of kilowatts." The position of our audience in the organization determines the tone or formality of our communication. What the audience already knows about this specific subject helps us decide the amount of background information to include, and the audience's values determine our appeal— should we emphasize company profits or stock dividends, company reputation in the industry or employee morale? The audience's reading habits help us organize our communication; for example, managers often read only the abstracts of technical reports while peers or those who share our specialty read the discussion sections thoroughly. The immediate and eventual use and the secondary audience of the communication help us decide how much detail to include as well as the tone to use; will our supervisor file our memo or pass it on to the vice-president? Finally, who will be affected by our suggestions determines the extent of them; if we recommend that the company close a branch, we should suggest how to relocate the branch employees.

# The Direction of a Communication

The direction of a communication also helps us analyze our audience and purpose. If our communication goes upward to a supervisor, we recommend action to a decision-maker. If our communication goes laterally across the hierarchy of an organization, we would suggest to a peer or equal but could not direct. However, we do instruct or direct employees. If we address a potential customer, we explain company procedures or recommend a product to someone outside the company. These types of communication can be called *upward* (to supervisors), *lateral* (to peers), *downward* (to subordinates), and *outward* (to customers, public interest groups, stockholders, the government, an so on). We must be aware of our own position in an organization as well as that of our audience. To give orders to a supervisor is just as disastrous as to command a peer, to write above an employee's educational level, or to overwhelm a potential customer with technical theory. We can make some generalizations about upward, lateral, downward, and outward communication to use in analyzing an audience.

## Upward Communication

Generally supervisors are often managers who are concerned with profits, major company procedures and changes, employee morale, new systems of organization or production, benefits or risks of new projects, and company progress and image in industry, research, and society. They need not only a thorough introduction to a communication and definitions of technical terms but also a brief summary of the whole communication to save reading time. Because managers and supervisors read many reports or abstracts of reports, our upward communication should be as concise and direct as possible.

## Lateral Communication

Peers inside the company may share our expertise in a field. In any case, they are concerned with how our ideas and projects affect their own research and jobs. Tone is perhaps the most important consideration in lateral communication; because we cannot invade a peer's territory, we suggest rather than command. If the audience of lateral communication shares our professional background, its members understand a technical vocabulary, symbols, and abbreviations, and need less background than do managers or supervisors. However, this audience is most critical because its members test both the logic and practical application of our ideas. Because they might apply our suggestions to their own research, they want a detailed account of our subject, concrete as well as theoretical.

## Downward Communication

Audiences of employees and technicians are most concerned with how to increase their productivity and thus their incomes. However, because they need to see how their jobs fit into the overall company operation, we should remember to explain as well as give directions. Although employees may not have the education we do, they might have more on-the-job experience. We need to recognize their practical knowlege.

## Outward Communication

If our communication goes outside the company, some of the techniques of upward communication apply. The members of this audience are busy with their own jobs or lives and want simplicity, defined terms, and a thorough introduction to our facts and ideas. However, this audience is most interested in how personal jobs, incomes, taxes, and life-styles will be affected by our ideas. Potential customers want to know about the cost, use, and durability of our products; the public wants information on how we will affect the economy or the environment; stockholders want to know how we are increasing stock dividends; the government needs to know how our organization is conforming to government regulations or affecting the national economy.

Some of these audiences can be combined. For example, if we write a letter to a manager in another company, our communication goes outward as well as upward; if we publish an article in a research journal, the communication goes outside our company but addresses peer professionals. In analyzing our audience, we must locate the direction of the communication but also apply the seven questions we discussed earlier to our specific audience.

# The Purpose of the Communication

The purpose of our communication is what we want the audience to think, feel, or do after reading it. While we communicate technical information for two general purposes, to inform and to persuade, we must analyze our specific reasons for communicating. Most technical communication conveys information; however, since we are experts in our fields, even though we write to a supervisor or a peer, *we* interpret our information—*we* decide what it means and how it can be used. What makes any communication dull is the absence of the thinking mind. The more we analyze our information, the more likely we are to influence our audience. Thorough audience and purpose analysis allows us to be as persuasive as possible. For example, our specific purpose might be to investigate for our supervisor which duplicating machines our company can lease within budget and space restrictions, and to recommend, in a report that the supervisor can pass on to the general manager, the best machine for our company's needs. We want our supervisor to be pleased with our thoroughness, be impressed with the logic of our recommendation, and be confident in passing on our report to his or her own boss. We might even want the company to put us in charge of purchasing in the future or in charge of teaching employees how to use the new machine. Analyzing our specific purpose also helps us write a clear, concise, useful communication.

# Rhetorical Patterns in Communications

We use rhetorical patterns to organize our communications and to convey our facts and ideas. Even before we begin to write, we use rhetorical concepts to organize our thinking. The audience and purpose of the communication determine which rhetorical pattern to choose. All letters, memos, reports, phone calls, interviews, and articles are made up of a series or combination of rhetorical patterns. We choose them according to the questions we ask ourselves while writing and the answers our audience seeks while reading our communication. The patterns we use most often and some of the questions we ask and answer follow:

**Definition:** What distinguishes this subject from similar ones? What does the subject mean in this situation? What is a synonym or more common term for the subject? What is the class or major group to which this subject belongs?

**Description:** What does this subject look like? What are its dimensions? What is it made of? What are its parts and how are they attached?

**Comparison-Contrast:** Are these subjects considered similar or different? What are the similarities among or the differences between them? What are the characteristics they share? How many do they share?

**Causal Analysis:** What caused this occurrence? Were there one or many causes? What will happen if this occurrence does take place? Will there be one or many effects?

**Classification-Partition:** In what categories do these subjects belong? What is the basis of their grouping? Can the parts of this unit be grouped according to a basis?

**Narrative:** What happened? When did it happen? Who observed or participated in the event? How did the event begin and end? How can it be retold or recreated in words?

**Process Description:** How does the subject work? What does it look like in action? What are the functions of all the parts? What causes the action of the parts? What part moves first, second, last? Does the subject operate according to a basic scientific principle or law?

**Procedure:** What is the goal of this operation? How should the people involved cooperate? What are the options or alternatives in completing the operation?

**Instructions:** How should this subject be operated? How should it be installed? Repaired? What steps are necessary? What steps are optional? What precautions should be taken?

**Specifications:** What are the requirements of the proposed project? What materials and methods have to be used? What legal requirements have to be met?

**Argument:** What doubt or conflict exists about the subject? What specific observations back this generalization? What are the reasons for this inference? Is the subject logically valid and true?

**Persuasion:** What are the facts, opinions, and inferences in our argument? What experts believe this proposition? Is this proposition ethical? How can the proposition be refuted?

We share these questions and their answers with our audience when we explore a subject and write a communication. We use one basic rhetorical pattern or a series or combination of patterns. For example, when we present a new idea, we might define our terms, compare and contrast the new and the old idea, describe how the idea will work, and analyze its efffect on various groups of people. If we know our audience and our purpose, we can choose appropriate rhetorical patterns. By knowing when and how to use the patterns, by learning what to include in them, and by knowing how they will affect or influence our audience, we can meet the requirements of any communication. We can explain any fact. We can convey and sell any idea. Our thinking becomes clear, and we can communicate effectively.

# Stylistic Analysis of Communications

From the first words of any communication, sentence structure, vocabulary, tone, length of sentences and paragraphs, and even titles and subheads are determined by the audience and purpose of the communication. We make all these stylistic choices according to the background and concerns of our audience, the level of the audience or the direction of the communication, and the overall purpose of the communication. Our choices ensure that our audience will know immediately why we are writing the communication and will be able to understand it. We can see what these choices are specifically and how to design a communication by looking first at some examples from magazine or journal articles and then at some examples from technology and industry.

## Journal Articles

**Sample One.**   The title of any communication should indicate the purpose and tone (the writer's attitude toward the subject) of the communication. The following paragraphs appeared in the beginning of an article entitled, "How Television Helps Johnny Read." In this title the writer promises to explain how television promotes literacy and indicates that the article will be light and perhaps amusing. Because the writer refers to the stereotype "Johnny," we also sense that the article has a general appeal. Does the article keep the promises made in the title?

> Can television encourage people to read more? Can it even help them read better?
>
> Those sound like two strange questions. They appear to be based on an upside-down assumption—like asking whether rain can help people get dry. For at least a quarter-century, many educators and millions of parents have assumed that TV is the natural enemy of reading, particularly among children. The connection was felt to be self-evident: the more you watch TV, the less you read and the dumber you get.
>
> But it may not be so. This is what this report is about.

Because the appeal of the article is indeed broad, the writer asks questions in the first paragraph to provoke the audience's interest. Since anyone interested in children, literacy, or television might read the article, the writer chooses simple, familiar words. Although the middle paragraph is long, it appears between two short paragraphs which makes the passage *look* readable. The tone is matter-of-fact, almost casual. Thus when analyzing his audience and purpose, the writer decided that he had to capture his audience's attention with an interesting, provocative opening. He might have opened also with a familiar quotation, an anecdote, a bold statement, or even a joke. He had to keep the audience's attention with short sentences and paragraphs, and his information had to be accessible and interesting to all.

**Sample Two.**   The next passage appeared in a magazine for an audience with practical experience rather than formal education in mechanics. The title of the article, "We Test the 1977 Plymouth," promises that the writer will address the audience directly and report clearly and simply the results of the tests. Because the audience's experience and level of technical education might vary widely, the vocabulary is not challenging; in fact, the writer uses slang. Because sentences and paragraphs are short, the audience has a visual "break" while reading. The purpose of the article is clear—to show how and why, despite predictions, one car is selling well. Notice that the writer also establishes a rapport with his readers:

> When the 1976 model year started, nobody expected the big cars to do much in terms of sales. The little jobs were going to be the heroes.
>
> Well, it didn't turn out that way. It has been the big cars that have sold well while the little fellows had all kinds of production cutbacks.

Which brings us to the test car at hand, Plymouth's Gran Fury Brougham two-door hardtop. Just call it the standard-size Plymouth for short. This car first came out as a 1975 and remains essentially unchanged in 1977. Except car-buying preferences have changed. When we tested this one as a '75, times were hardly auspicious for a rig of the Gran Fury's size. And sales weren't good.

But as of this writing, the showrooms are buzzing with people clamoring for those big, comfy Detroit chariots. So the first '77 test that we bring you from Motor City is a car that finally is likely to go someplace—the Plymouth Gran Fury. . . .

Standard power plant for the Brougham series is the lean-burn 400-cu.-in. V8. Several sensors and a small computer located next to the 4-bbl. carb control spark-plug timing electronically. It's called lean-burn because it uses a higher numerical ratio of air to fuel. Conventional mills run a 16:1 air/fuel ratio, lean-burn 18 or 20 lbs. of air to 1 of fuel. Leaner air/fuel ratios produce lower emissions and increase fuel economy. Starting and drivability are also helped.

The writer mixes technical detail with a conversational tone.

**Sample Three.** The next passage appeared in a magazine for college professors. The title of the article is "Hardness of Soul, Sclerosis of Imagination." What promises does the title make to the audience? How are the vocabulary, tone, and sentence and paragraph length appropriate to this audience? What is the purpose of the article?

He [a professor] is, furthermore, talking about students, including the best and the brightest. Today's challenge to higher education is that so many young men and women arrive on the educational scene *already* possessed by such "hardness of soul" and "sclerosis of imagination" that it almost amounts to invincible ignorance. They are possessed by a hard-nosed, yet quaveringly anxious, careerism, bereft of any real capacity for either moral passion or aesthetic delight. The theological term for their malady is "concupiscence," that Faustian condition in which the God-presumptive soul—hungry to bring all reality under its hegemony, but everlastingly insecure about some outstanding, elusive portion of it—pursues its joyless, predatory career through the world.

The purpose of the article is to convince the audience that today's college student seeks an education that guarantees a job rather than provokes the imagination. Because the audience is well educated, the writer alludes to literary characters and uses philosophical, theological, and medical terms. The vocabulary would discourage a more general audience. The sentences and paragraphs are long, and the tone is formal. The writer chose not only content but also style appropriate to college professors.

So far we have studied how writers of journal articles choose a style and content appropriate to various audiences and purposes. However, since in technical communication we address a particular *specialized* audience, we must sophisticate our study of style. We look next at some examples of on-the-job technical and business writing.

## Technical and Business Communications

*An Example of Upward Communication.*  The following example addresses an audience of decision-makers. A company hired an engineering firm to investigate the cost of renovating a building that the company wants to occupy. The following passage is part of the consulting engineering firm's report. Here the communication is directed upward to an audience of managers:

> The first [major area] concerns the condition of cooling coils, supports and casing. Although we did not physically inspect every air-handling unit internally, those that we did (six in total) were found to show signs of age, in that the fins were brittle and pitted and the casings and coil supports showed signs of rust build-up. These units have been in operation for approximately 18 years, are at the end of their useful life, and are unlikely to last for any extended period of time. The average life for air-handling equipment in New York City is about 15 years.
>
> Since in our opinion the ultimate need for replacement of air-handling equipment is only a question of time, the present change of ownership of the building and [the] scheduled occupancy presents a unique opportunity to replace existing air-handling equipment with new equipment of greater efficiency. We also recommend that at the same time the interior systems be changed to variable volume. The design of the systems will reflect the reduced lighting load of the selected lighting scheme. The total gain in efficiency achieved through these changes will result in considerable energy savings for the entire building.
>
> The estimated replacement cost for those systems serving the perimeter and interior spaces, induction and dual duct systems respectively, is $4,000,000. . . .

Again the communication addresses the audience's major concern—cost. The tone is formal, and the vocabulary appropriate for an educated but nontechnical audience. Because the writers do not have to capture the audience's attention, sentences and paragraphs are long but clearly written. The purpose of the report is to inform the company about the need for and cost of renovating the building and to persuade the company to take the best action.

*An Example of Lateral Communication.*  Because in lateral communication we address an audience of experts, peers in education and experience, we use technical terms and symbols. Our audience is concerned with theory as well as application. The following passage appeared in a technical paper for manufacturing engineers. Notice that the writer discusses both the theoretical basis for the developments he shares with the audience as well as the specific advantages to the field:

> There is considerable evidence in nature that demonstrates the ability of moving water to erode and transport solid materials. Many geologic formations are attributable to water action on soil and rock but the removal rates are often very slow.

Engineering advances in the design of pumps, seals, tubing, and nozzles have made it possible to extend the idea of liquid erosion of solids to uses where materials much harder than soil and sandstone are to be cut. Moreover, the higher velocities attained have fostered the use of water jets for mining and excavation work as well as for producing manufactured products. Steady flow pumping systems that produce pressures up to 60,000 psi (414 MN/m$^2$) and deliver jet velocities of approximately 2500 feet per second (762 mps) are now commercially available. With these new pumps, it is possible to cut materials of moderate strength and hardness.

There are a number of advantages which can be claimed for high velocity liquid jet cutting when used in various applications. The removed solid material is usually entrained in the liquid and consequently the operation becomes dust-free, reducing hazards in mining and some fabrication work. Because jets of small size (as small as 0.005 inch [0.126mm] diam.) can be produced, wasted material from the kerf is minimized in a manufactured product. . . .

Because the writer also explains logically the results of the engineering advances, sentences and paragraphs are developed and the vocabulary specialized.

*An Example of Downward Communication.* Employees are often concerned with how to make their jobs easier and how to increase their salaries. Therefore, we address employees more directly than other audiences and may use a more informal tone. Although employees may not share our area of expertise, they usually know their jobs well. In the following passage, taken from a company manual, although the writer does not use a technical vocabulary, he does discuss specific tasks in detail. Because he gives precise, simple directions, the sentences and paragraphs are short. He addresses his audiences as "you," an informal approach:

All mail received (except registered) up to 5:15 P.M. will be dispatched that evening. Out-going registered mail received in the mail room any time prior to 4:30 P.M. on normal working days will be delivered to the post office with the registry book for dispatch. Registered mail received after 4:30 P.M. will normally be dispatched early the following morning, unless the sender indicates an urgency for evening dispatch. In this case, arrangements will be made for delivery to the post office on an overtime basis.

If you are not sure of the proper mailing class, note the contents on the lower left-hand corner of the envelope. The mail room will then classify it properly. Special handling instructions should be indicated on the envelope (air mail, special delivery, insured, parcel post, registered). Do not use window envelopes for registered mail.

The purpose of the communication is to instruct employees so they can be more efficient.

*An Example of Outward Communication.* Potential customers are interested in the costs and abilities of our products. Although they may have some technical education and experience, they are more interested in the end results of our investigations, the product. The following passage appeared in a company catalog. The writers use a basic technical vocabulary to explain the functions of the products. The sentences and paragraphs are fully developed but not complex, and the tone is formal and objective. The audience is more interested in a clear choice than in a hard sell:

> Electronic field devices such as transmitters, transducers, positioners, and switch contacts are sometimes required to operate in an explosive atmosphere. The field devices are electrically wired to and operate with controllers and other instrumentation located in the non-explosive atmosphere of a control room. It is necessary in these situations to assure that an accidental electrical spark is unable to ignite the explosive atmosphere. This can be accomplished by encapsulating the field device and its wiring in an explosion-proof housing. However, a more economical method of preventing ignition is to limit the electrical energy entering the explosive atmosphere to a level that is far below that required for ignition. This energy level must be limited even in the event of an accidental short or open circuit.
>
> Fisher Controls offers a variety of intrinsic safety barriers that limit the electrical energy entering an explosive atmosphere. Each barrier is wired in series with the instrumentation signal wiring and is mounted where the wiring enters the explosive field atmosphere. The barriers limit the energy level by shunting excessive current to earth ground and by regulating voltage with zener diodes. Our complete line of barriers will operate with transmitters, transducers, positioners, field contacts, thermocouples, and resistance elements. . . .

The layperson considering buying these safety barriers could understand the description and choose the best product for his or her situation.

The content of our technical communications addresses our audience's interests and concerns. The style of the communication—its vocabulary, sentence structure, title, tone, and sentence and paragraph structure—is determined by our audience, direction, and purpose. We choose our words according to our audience's experience and education, positions inside or outside the company, and use of the technical communication; according to the direction of the communication—upward, lateral, downward, or outward—; and according to our specific purpose.

# Summary

In technical communication, we address a particular audience with a special interest or knowledge. We present and recommend facts and ideas in a clear, concise, and logical way. Before we write our communication, we analyze our audience's background, experience, interests, values, reading habits, knowledge of the specific subject, and position in the company as well as the needs and interests of any secondary audiences. Technical communication can travel upward to supervisors, laterally to peers, downward to employees, and outward to potential customers, the public, the government, stockholders, and so on. We also analyze the specific purpose of our communication—what we want our audience to think, feel, or do after reading it.

We organize the facts and ideas in our communication according to rhetorical patterns which help us answer the questions we and our audience ask about our subject. Audience analysis also enables us to choose an appropriate style and content for our communication. We choose our words, title, tone, sentence and paragraph stucture, and approach to our subject according to our audience and purpose.

We keep these techniques of audience and purpose analysis in mind as we study the rhetorical patterns used in technical communications in the next chapters.

# Reading
## to Analyze and Discuss

# I

## A
## Cosmos 954: An Ugly Death

No cause for panic, said the U.S. National Security Adviser Zbigniew Brzezinski. It had merely been "a space age difficulty . . . There is no danger."

The little difficulty that Brzezinski so soothingly soft-pedaled was the fiery return to earth of Cosmos 954—a Soviet spy-in-the-sky satellite carrying a nuclear reactor to power its ocean-scanning radar and radio circuitry. The craft crashed into the atmosphere over a remote Canadian wilderness area last week, apparently emitting strong radiation. American space scientists admitted that if the satellite had failed one pass later in its decaying orbit, it would have plunged toward earth near New York City—at the height of the morning rush hour.

The event gave the public a rare glimpse, fascinating and fearsome, of the two superpowers tiptoeing through a two-step diplomatic dance. It also offered a shocking reminder of the masses of hazardous hardware now orbiting through our heavens. . . .

# B
## The Unscheduled Return
## of Cosmos 954

Cosmos 954's reactor, according to one U.S. analysis, was estimated to occupy a little less than a cubic meter (although finding even a piece of it might enable a much more accurate calculation). Its 100 pounds of uranium 235 could imply a power output in the range of tens of hundreds of kilowatts. Any recovered scraps of the device would be informative, and U.S. officials early this week were hoping for what could be an intelligence bonanza, revealing details of the state of Soviet reactor design, metallurgy and other technologies including the radar system.

. . . The last U.S. nuclear fission reactor acknowledged to have been used for spacecraft power was launched in 1965, carrying 25 pounds of U 235 and designed to burn up and disperse in a high orbit. Since that time, however, a number of earth-orbiting U.S. satellites (as well as deep-space probes) have carried radioisotope thermal generators (RTG's) which work by the passive thermoelectric conversion of heat from their fuel, typically plutonium. The President's FY 1979 budget request for the Department of Energy, issued only a day before the Cosmos incident, also contains several million dollars to begin development of a 10- to 100-kilowatt nuclear reactor for spacecraft use. (A department official says that, unlike the case of Cosmos 954, which apparently failed to ascend as planned to a higher, longer-lived orbit, any reactor aboard a U.S. spacecraft would not be turned on until it was safely in high orbit. The initial fuel load of U 235 is considered a relatively minor hazard compared with the by-products, such as

**16**

strontium and cesium, that accumulate during its operation.) Funding has also been requested for improved-efficiency RTG's, using a "dynamic" heat-conversion system rather than direct thermoelectrics. . . .

# C
# The Satellite That Invaded a Campsite

The six men had originally planned to stay in the wilderness for fifteen months. They would travel from west to east, starting in the Yukon and moving across the vast landscape into the Northwest Territories: up the Mackenzie River to Great Slave Lake. They would winter in the Thelon Game Sanctuary, and when the thaw came to that brutally cold land, they would take their canoes downriver and into Hudson Bay. To anybody who has done any canoeing at all, the ambition of the trip is staggering. [Not a trip, in fact, at all, but an expedition and an adventure in the true and certain sense of those overworked words.]

But their trip turned into something much more than an adventure. They found things they weren't looking for and wished they hadn't found. [Things that changed their lives forever and, in a sense, the rest of the world's also.] For seven months they were free; then the world literally crashed in on them near Warden's Grove. The adventure turned dark: man and technology exploded into their plans, their wilderness. . . .

Things were going well until John Mordhorst and Mike Mobley went out on a two-day dogsled expedition. At a frozen riverbed, they came across an odd-looking metal object, which they stopped to examine. When they returned to camp, they learned from the others (who had heard the news by radio) about the fall of a Russian satellite that carried a nuclear reactor. There was a good chance they had discovered a chunk of it and had been contaminated with radiation. . . .

# D
# Handwriting on the Sky

The world has been living through a dress rehearsal for nuclear Armageddon. To say that is not to join the loud alarmist chorus about the Soviet spy satellite whose orbit decayed into the Canadian northwoods, its nuclear power plant scattered through the wilds.

Let us accept the complacent words of Zbigniew Brzezinski that this was just a "space age difficulty." Let us put out of our minds the fact that, now that Cosmos 954 is down in radiant smithereens, there are still 938 other satellites whirling around in the heavens. The world won't end this way and it would have survived even if, on the next pass, the Cosmos had come to its fiery end on a New York street in the rush hour.

What is frightening about this incident are the things called to mind by the system to which Cosmos 954 belonged. The system is called "defense," here and in Russia. Now and again we get a glimpse of its awful capabilities—satellites that not only spy on one another and on the earth below but can knock each other out and trigger the nuclear weapons of both superpowers on the command of a fallible mortal still on the ground. . . .

# Questions for Discussion

1. What are the stylistic differences between the openings of the four articles on Cosmos 954? What promises are made in the titles? How much technical detail appears in the articles? Are words used that carry certain connotations or associations? How are the openings designed to capture the audience's attention?

2. What are the specific purposes of each article? What does the writer want the audience to think, feel, or do? Do the writers represent a certain political, social, or economic philosophy? How many facts appear in the articles? Opinions?

3. Describe the audiences addressed in the articles. In what magazines or journals could these articles have appeared? What are the education, experience, knowledge, values, and reading habits of the audiences?

# Exercises and Assignments

**1. Problem:**   Your technical supervisor has asked you to investigate current research on a particular subject. In your report you are to include an audience analysis of each article you discuss. You realize that in your audience analysis you must also comment on the direction and purpose of the publication. You need to find evidence or support for your audience analysis through stylistic analysis.

Assignment:

1. Choose a subject or area of research with which you are familiar.
2. Find three articles or reports on that subject or area. You might find articles listed in the *Reader's Guide to Periodicals* or in a bibliography in your technical field.
3. Analyze the audience, direction, and purpose of the article.
4. Find specific evidence for your analysis: vocabulary, sentence and paragraph structure and length, tone, title, opening paragraph, and so on.
5. Report on your findings in steps 3 and 4 in a short communication to your supervisor.

**2. Problem:**   You are a manager of a large department store in downtown Chicago. The store also has three suburban locations.

Assignment:

List what you would want to know about your audience if you had to write the following communications:

1. A list of instructions for your employees on how to display merchandise.
2. A report to the supervisor in charge of personnel which makes a request for an additional clerk.
3. A memo to the managers of all three suburban branches which describes a new way to schedule employee hours. You are now trying this new method in your store.
4. A public relations release to the local media announcing the opening of a fourth suburban branch.
5. A flyer to be sent through the mail announcing your Washington's Birthday sale.

**3. Problem:**   You have been asked to investigate and report on a certain subject. You must discuss this subject with a variety of audiences.

Assignment:

1. Give yourself a title and position within an oganization. Your organization can be professional or educational.

2. Choose a subject with which you are familiar or about which you would be willing to do some research.
3. Write the opening paragraphs and titles of five communications on the same subject for five different audiences. Your five audiences are: the high school student; the college student; the peer professionals or students on your level in the same major; a supervisor, teacher, or boss; a general audience.
4. Preceding each title and opening, describe your audiences in detail and state the purpose and direction of your communication.

**4. Problem:** You are employed in the public relations office of a major company. Your most important job is to release technical information to the general public through the media and to employees through the company magazine.

### Assignment:

1. Find a report or article directed upward, to a supervisor or to an audience with a great deal of technical knowledge. Be sure to choose an article in your major field or one which discusses a subject about which you know a great deal.
2. Rewrite the opening paragraphs and title of the article or report as if the communication addressed the general public.
3. Rewrite the opening paragraphs and title of the article or report as if the communication addressed company employees.

**5. Problem:** You manage a claims office for a major insurance company. There are eight claims offices in the country, all of which report to or are under the supervision of a home office. You have an idea about a quicker, more efficient way of processing claims which you would like to try out in your office. The process involves combining two forms presently used. One form asks for the details of the claim: names, dates, type of insurance, etc. The other form describes the action taken: claim denied, claim approved and amount paid, further investigation recommended, and so on. You would like to use one form with several carbon duplicates attached. The first copy, with the details of the claim, would be completed and filed. The second copy would be completed after the claim is reviewed, but would have all the original information duplicated on it. You are writing to the home office to recommend the new combined form. You are also writing to one of the other claims offices, the one closest to you, to describe your idea and to ask their opinion about the combined form. You also need to write a memo to the clerks and typists in your office to ask for any further suggestions about the format of the combined form.

**Comments:** You may add details about the form if you wish.

Assignments:

1. List the questions you need to ask about each audience before you do the writing called for in Nos. 2, 3, and 4 below.
2. Write the opening paragraphs for the upward communication.
3. Write the opening paragraphs for the lateral communication.
4. Write the opening paragraphs for the downward communication.

# 2

# Gathering and Organizing Technical Information

*Gathering information from secondary sources.*
*Taking notes.*
*Summarizing objectively or critically.*
*Gathering information from primary sources.*
*Creating an outline from notes.*
*Reading and analyzing the research done by others.*

While thorough audience analysis is one key to successful technical communication, two other techniques—gathering and organizing information—are also basic to every communication. The rhetorical patterns we study in this book help us present information for specific purposes. But first we must collect that information whether we research our subject in the library or in the laboratory. Then we organize our information and decide which rhetorical patterns to use. In this chapter, we study the techniques of gathering and organizing information, techniques that make our writing task manageable no matter what the size of the project.

When we first choose or are assigned a project, we engage in some sort of research. We may collect information from books or periodicals, question other people, perform experiments, or such. During our research, we define and redefine our subject as we learn more about it; we may limit the scope of our subject so we can study one aspect in depth.

For example, if our supervisor asks us to investigate which duplicating machine our company should lease, we would limit our research to models that best meet the company's specific needs and budget. If the company needs a duplicating machine that can make back-to-back copies and reduce originals and that costs under $350 a month, we would first skim the sales material on all duplicating machines to eliminate those that do not meet all these criteria. Then we would interview sales representatives, read business magazines, and study the annual reports of the companies that lease the machines that do meet our criteria.

Whether we collect our information from written material or from personal interviews, we take notes on that information. We record the information word for word or summarize it in our own words. When we summarize other people's ideas, we can do so objectively or critically. For example, we might quote exactly the prices and features of the duplicating machines but critique the sales material or the performance record of the companies.

Finally, we gather all our notes and arrange them in the order we want to present them in our communication. This order is reflected in an outline that helps us write a logical, complete communication. Whether our project is a large one such as investigating duplicating machines or a small one such as writing a letter of complaint, we gather and organize information before we write our communication. These "prewriting" tasks, along with a thorough audience and purpose analysis, help us write a concise and yet complete communication.

# Gathering Information from Secondary Sources

A secondary source of information is a work that contains someone else's ideas, thoughts, and interpretations. Books, periodicals, reports, dictionaries, encyclopedias, booklets, and such are all secondary sources of information. They are the sources we turn to first to provide the background for our project. To conduct original research we have to know what other people have said about our subject. For example, the sales material about duplicating machines, a secondary source of information, would provide us with the background for interviewing sales representatives, our original research. Now we discuss the ways to use secondary sources and the sequence in which to use them.

## General Guides to Secondary Sources

If we are unfamiliar with our topic, the best place to start our research is with guides to secondary sources. These general guides help us find major sources which in turn lead us to more specific sources. General guides list and often describe or annotate sources according to subject or field. By starting with a general guide to secondary sources, we can see how much information is available on our subject as well as which sources are relevant. Although general guides are gathered in the reference section of most libraries, the following guides are particularly useful to scientists, technologists, and business people:

E. T. Coman, Jr., *Sources of Business Information. Encyclopedia of Business Information Sources.*
Frances B. Jenkins, *Science Reference Sources.*
H. W. Johnson, Ed., *How to Use the Business Library—with Sources of Business Information.*
Robert W. Murphey, *How and Where to Look It Up—A Guide to Standard Sources of Information.*
Eugene P. Sheehy, Ed., *Guide to Reference Books.*
Albert J. Walford, Ed., *Guide to Reference Material*, 3 vols.
Constance M. Winchell, *Guide to Reference Books.*

Sheehy, one of the most useful guides for technologists, lists other general guides, bibliographies, indexes, handbooks, dictionaries, directories, periodicals, abstract journals, encyclopedias, and yearbooks for specific fields.

## Guides to Books and Periodicals

Within such general guides as Sheehy, we find more specific guides to types of secondary sources, such as *Engineering Index.* Each specific field or area has one or more indexes or abstract journals that list and sometimes describe books and articles on any topic in that field. These more specific guides are our next step in basic research or the first step if we are somewhat familiar with our topic already. Some typical entries from indexes or abstract journals in special fields appear below. Coding, style, and abbreviations are particular to each index and explained in the preface to each index:

From *Applied Science and Technology Index* (under the subject of fuel economy):
Aerodynamics; the new design input: R. J. Fosdick. il Automot Ind 158:22–9 Mr '78.
Aerothermodynamic performance of a variable nozzle power turbine stage for an automobile gas turbine. E. H. Raxinsky and W. R. Kuziak, jr. bibl diags J Eng Power 99: 587–92 O'77.

From *Engineering Index* (under the subject of accelerators):
NEXT GENERATION OF ACCELERATORS. An examination is made of the physics issues which go into setting machine parameters, and some of the features of the design of next generation electron and proton machines. Richter, B. (Stanford Univ, Calif). IEEE Trans Nucl Sci v NS-26 n 3 pt 2 Jun 1979, Proc of the Part Accel Conf on Accel Eng and Technol, 8th, San Francisco, Calif, Mar 12–14 1979 p 4274–4276.

From *Business Periodicals Index* (under the subject of factories):
Factory builders find slim pickings. il Bus W p51–2 Jl 24 '78
High cost of making an industrial move. R. I. Wilson and others. il Mgt R 67:31–4 Ag '78.

If we are researching fuel economy in such guides as Sheehy, we would find under the general topic of technology the *Applied Science and Technology Index.* When we looked up our subject, fuel economy, in the *Index,* we would find a list of sources that deal specifically with it. One such article appears in the March 1978, volume 158, issue of the periodical *Automotive Industries* on page 22, and the other in the October 1977 issue of *Journal of Engineering Power* on page 587. The entry on the article in *Engineering Power* states that the article contains a bibliography, which might in turn lead us to other sources.

If we are doing research on accelerators, the article from *IEEE Transactions on Nuclear Science* would be relevant if we were interested in design. Finally, if we are doing research on factories, we could find information in a general article in *Business Week* and a more specific article in *Management Review.*

By now it should be clear that research in secondary sources is a matter of going from the general to the specific both in terms of sources and topics. We start with general guides, turn to indexes or abstract journals in specific fields, and end with specific books and articles on our topic. We start with a general topic, such as fuel, and find that a more applicable or manageable topic is fuel economy or even new aerodynamic designs for fuel economy.

## Guides to Other Secondary Sources

While books and periodicals are often our main sources of specific information, we can also find information in newspapers, government publications, reports, bulletins, and brochures.

Major newspapers such as *The New York Times* and *The Wall Street Journal* publish an index of articles that appear in that specific newspaper. The articles are listed by subject as well as person or organization.

Government agencies publish pamphlets and booklets on almost any subject. These are indexed in the *Monthly Catalog of United States Government Publications* from the U.S. Superintendent of Documents as well as in such guides as Ellen Jackson's *Subject Guide to Major United States Government Publications.*

Reports, bulletins, and brochures are indexed in such sources as Raphael Alexander's *Business Pamphlets and Information Sources* and the *Gebbie House Magazine Directory.*

# Locating Secondary Sources

## The Card Catalog

Once we have found periodicals, books, or other materials that contain information we are looking for, we must locate those sources. If our library has them, they are listed in the card catalog by title, author, and subject. (In some libraries periodicals are listed by title in separate catalogs.) The card catalog can also be a thorough source of works on a subject if we use a major library. A sample listing from the card catalog follows:

```
B67      Bronowski, Jacob, 1908-1974
.B7          A sense of the future: essays in
1977     natural philosophy / J. Bronowski;
         selected and edited by Piero.E. Ariotti, in
         collaboration with Rita Bronowski.
         Cambridge: MIT Press, c 1977
             x, 286 p. ; 24 cm.
             Includes bibliographical references and
         index.
             1. Science-Philosophy-Addresses,
         essays, lectures. 2. Time perspective-
         Addresses, essays, lectures.
         I. Title
NPotU    13 APR   78 MW      ZOMMdc    77-9292
```

The book would be listed under the title *A Sense of the Future* and under the two subjects at the bottom of the card. The number in the left-hand corner is the call number which indicates the book location in the library; the information below the publisher and copyright date tells that the preface is 10 (x) pages long and the book 286 pages long and 24 centimeters in size. The librarian uses the code at the bottom of the card to reorder the book.

While the card catalog may be the most familiar spot in the

library, we should remember that the listings in the card catalog can also serve as guides to sources. Bronowski has written other books on similar subjects that would be listed before or after this one. If we look under the subject headings listed at the bottom of the card, we might find other sources relevant to our topic. Bronowski's own book contains a bibliography of sources. Even when we reach the point of locating our sources in the library, the card catalog may give us clues to other sources.

### Interlibrary Loan

If our library does not have the book or periodical we need, the librarian can order it through a search and borrowing system called interlibrary loan. If the source is located in another library in the area, we can usually have the source within a few days. However, if the book is located in another state, the search and processing take time, a very practical reason to start our research project well in advance of the due date.

## Computer-Assisted Searching

A new aid to research is now available in some libraries—computer-assisted information retrieval. Hundreds of data bases or computer readable bibliographical sources store the information listed in indexes in many specialized fields. For example, the data base COMPENDEX stores past listings of *Engineering Index.* Data bases often have listings available before the indexes themselves appear in print. With the aid of the information-retrieval librarian, we can "call" for all the entries on a specific subject or key word, such as "fuel economy," and receive a computer printout of the bibliography resulting from the search. In other words, we can let the computer compile a bibliography of books and periodicals on our subject. However, before we can call for a computer-assisted search, we need to define our subject precisely, a process that usually takes some research on our part.

## Taking Notes from Secondary Sources

Once we locate our secondary sources and begin to read them, we need to record information relevant to our project. We can record direct quotations, word-for-word statements from the source, or summarize the main ideas in the source. Although we should be brief in our note taking, we should try to record as much as or more information than we will need in our communication so that we do not have to stop writing and return to a source. One of the best methods for taking notes from secondary sources is a system of note cards. On note cards we can compile both a working bibliography and the information we need.

# A Working Bibliography

Before we even begin to read a secondary source, we should record on a note card the author, title, name and location of publisher, volume and number if a periodical, and page numbers of the source. When we finish our research, we have a list of sources on cards that we can use to form a bibliography or write our footnotes. The 3-by-5- or 4-by-6-inch cards are easy to arrange in alphabetical order. A sample bibliography card follows:

> J. Bronowski, *A Sense of the Future: Essays in Natural Philosophy*, Piero E. Ariotti, Ed., Rita Bronowski, Collab. (Cambridge: MIT Press, 1977), pp. 255–256.

# Direct Quotations

If the source's exact wording is so striking or precise that its meaning would be lost in a summary, we should quote the source directly. We can quote the entire statement or just significant parts. Making clear in our notes direct quotations helps us avoid plagiarism or representing someone else's words or ideas as our own. Ellipses, three marks (. . .), indicate omissions, and square brackets ([ ]) our own additions or clarifications in a quotation. Since we should always introduce or attribute a quotation, we indicate on our note card what preceded the statement.

If we record a direct quotation from Bronowski, our note card would appear as follows:

> Bronowski, *Sense*                                                   1
>     After explaining that science gives meaning and order to
>     facts, J. Bronowski says: "Science is the human activity
>     of finding an order in nature by organizing the
>     scattered meaningless facts under universal concepts."
>                                                         p. 255

We use one side of the card, and we "introduce" or set up our quotation. At the top of the card, we key the source by name and by short title and number all the cards from this source; at the bottom we record the page number on which the quotation appears.

If we were to omit words and add comments to the quotation, we would record the statement as follows:

> Science is . . . finding an order in nature by organizing the scattered meaningless facts under universal [often abstract] concepts.

# Summarizing

When we summarize, we restate briefly a source's ideas in our own words. Our restatement captures the essence of the source's idea but is much more concise. While we might quote directly a significant word or two, summarizing helps us understand as well as record the source's ideas because we filter them thorugh our own minds. Here is a paragraph from Bronowski's book, followed by a summary of that paragraph:

> Consider now a very abstract concept: the concept of gravity. Newton invented it, in the form of a universal law of inverse squares, in order to organize a body of findings each of which was itself already abstract. For example, they included the laws found by Kepler, that each planet moves in an ellipse with the sun at one focus, and how fast each planet moves in its ellipse. These laws, in turn, depended on the ingenuity of Copernicus in seeing that the paths of the planets and of the earth make a simple pattern, a unity which is meaningful, if they are all looked at from the sun. And Copernicus, we must remember, was heir to centuries of astronomical speculation which, though we now think it out of date, did establish the profound idea that the paths of the stars are evidence of some deeper organized mechanism in nature.

---

```
Bronowski, Sense                                          2
Newton's concept of gravity was based on Kepler's theory of
the speed of a planet in its ellipse and Copernicus's theory
of the unified patterns of planets around the sun. In turn,
Copernicus's theory was based on past ideas that nature was
an "organized mechanism."
                                                    p. 255
```

---

Although we restate Bronowski's idea in our own work, we would still credit him with his thoughts on the concept of gravity in a footnote.

## An Objective Summary

While frequently we summarize ideas on our note cards, sometimes we include a summary of an entire section, chapter, or article in our communication. We share that formal objective summary with our audience. For example, our supervisor may ask us to summarize recent articles on a specific subject, or we may review a book for a professional journal. In any case, an objective summary is a condensation of the main ideas, structure, and tone of a work.

One method of writing a summary is to underline or record the key words or phrases in the original and then restate them in our own words but in the style, sequence, and proportions of the original. Since in an objective summary we represent the writer, we do not include our own thoughts. In an objective summary, we tell *what* the original source has said; we represent the original to our own audience.

## A Critical Summary

A critical summary, or critique, not only summarizes *what* the original source said but also evaluates *how* well the source said it. In a critical summary, we review the writer's style, evidence, approach, development, and anything else that pertains to how well he or she presents the ideas or content. A critical summary then contains a formal condensation of content along with an evaluation of style. In a critical summary, we must represent the essence of the source, but our own judgments are as important as our summary of content.

# Gathering Information from Primary Sources

A primary source of information is one that is original to the writer and that has not been subjected to the interpretation of others. When we gather information from primary sources, we interview people, send out questionnaires, conduct laboratory experiments, observe on-the-job situations, or use any other logical methods to gather original information. In effect, we become the source of information; our personal observations or our interpretations of the responses to our questions comprise the evidence or data we use in our communication. Primary research is done after gathering information from secondary sources.

## Interviews

The audiences and purposes of interviews vary greatly. We interview potential employees to assess their abilities; we interview customers to improve products and services; we interview employees to evaluate their performance; we interview experts to gather facts. While we cannot study each of these types of interview here, we can discuss two techniques that apply to most successful information-gathering interviews.

*Planning the Interview.*   Much of the success of an interview is determined before it even takes place. When we interview to gather information or conduct primary research, we must preplan the interview to guarantee we will get the information we need. For example, if we interview the sales representative of a duplicating-machine company, we need to do enough research on the company before we conduct the interview so we can ask appropriate questions. We might read the company's sales material and annual report and prepare a list of key questions to ask the representative. While other questions might occur to us as we go along, we have a reminder of what questions are essential to get the information we need. Research in secondary sources, along with a clear idea of the purpose of the interview, help us form the questions.

The purpose statement and key questions of the interview with the duplicating-machine company's sales representative might appear as follows:

**Background:** Alman Company's X-112 Duplicating Machine. Has back-to-back copy feature and original reduction feature. Leases for $250 per month including routine service calls every

3 months and 24-hour repair service. Company has recently merged with Dup-Print Company of Detroit and shows profit for the last 3 years.

**Purpose of Interview:** Determine details on how well X-112 meets company criteria and determine additional features of X-112. Also set up demonstration time for X-112.

**Key Questions:**

1. Is leasing cost subject to change, and if so, how much notice given to leasees? contract for how long? renewable for how long?
2. Will X-112 make back-to-back copies on any size paper? time to set and adjust machine for back-to-back? time to run machine for back-to-back copies? repair record for back-to-back feature?
3. How many grades or percentages of reduction for originals? time to set and adjust machine for reduction? any size paper? quality of reductions?
4. Twenty-four hour service a firm guarantee? Fridays? weekends? loan of other machine for major repairs?
5. Demonstration to determine speed and overall copy quality? when and where?

Although we might find answers for some of these key questions in sales material, asking the sales representative lets us compare his or her personal response with the information printed in the sales material.

*Listening Effectively during the Interview.* While we can preplan the interview by outlining essential questions, the purpose of the interview is to get answers. We have to listen effectively as well as record information we hear. The most effective way to listen during an information-gathering interview is to listen to understand, not to evaluate. We can assess the information we gather after the interview. During the interview, we have to listen without judging the person offering the information or the information itself. One way we can listen effectively is to repeat or summarize what the person says either immediately after it is said or at the end of the interview. This summary helps us check whether we have heard and recorded the information correctly; it also reminds the person we are interviewing of any information he or she may have forgotten to give us.

Interviewing is a complex skill we develop with practice. Preparing key questions before the interview and summarizing the information we hear and record help us gather the information we need to complete our project.

## Questionnaires

If we need to ask many people the same questions to gather our information, we might send them questionnaires. Although there is no guarantee that everyone will respond to the questionnaire, and we cannot ask on-the-spot questions as we can in a personal interview, responses to questionnaires are easier to tabulate and compare. We can ask the same question of a random sample of whatever audience with which we are concerned: customers, employees, peers, or such. The respondents may answer more freely or honestly than they would in a face-to-face interview; however, our questions have to be well presented, since we cannot clarify them as the respondents read them. Questionnaires are a primary source of information, since again we as writers ask the questions and interpret the responses.

*Closed Questions.* Most questionnaires are based on closed questions. Closed questions offer the respondent a choice of answers to check or circle. For example, in a questionnaire sent to sales representatives of duplicating-machine companies, a closed question would be as follows:

1. Please check below the reduction features of your duplicating machine.

   _____ a. 10% reduction only
   _____ b. 10–20% reduction
   _____ c. 10–20–30% reduction
   _____ d. 10–20–30–40% reduction

By giving respondents answers to choose from, we can easily tabulate and compare their responses. Most questionnaires are based on closed questions, since the value of a questionnaire is in the number of people it can reach and the ease in assessing their answers. Our questionnaire to sales representatives of duplicating-machine companies would help us select the ones we want to interview.

The questions we ask in a questionnaire follow in logical order, one leading into the next. If we ask the simple questions first, our respondents will be more willing to answer a few complex questions later. Questions should be objective and clear and cover one topic per question. If we mail our questionnaire, it should be accompanied by a personal cover letter that encourages our audience to respond. We should state in the letter when we need their response and enclose a self-addressed stamped envelope. Most people will answer a questionnaire if we can convince them that the survey will benefit them in some way. For example, we can stress that their answers will help us furnish them with a better product, improve service to them, or just allow them to express their opinions confidentially.

## Experiments and Observations

Our own observations, especially those we record when conducting a formal experiment, are also primary sources of information. When we present our observations, we usually state the purpose of the experiment, our methods, our detailed observations, the results of the experiment, and our conclusions based on these results. We must be as thorough and objective in our presentation as we were in our experiment, and we must report what failed as well as what succeeded.

# Outlining

Once we have gathered all our information from primary and secondary sources, we can organize and outline our communication. An outline should evolve naturally from our information, rather than be imposed upon it. The best outline is a "created" one.

## Creating an Outline

Let us suppose that our supervisor has asked us to investigate and recommend duplicating machines with the same criteria we used earlier in the chapter. We have now read the sales material and interviewed sales representatives as well as researched annual reports and business magazines. We have recorded our information from primary and secondary sources on note cards. To create an outline we need to "see" our notes and to place them in an order appropriate to our audience and purpose. The first step is simply to list the main facts and ideas from our notes without worrying about order. They might appear as follows:

Alman X-112 $250 a month to lease
X-112 routine service every 3 months
X-112 24-hour repair service (guaranteed except for Fridays and
    weekends)
Conmo C-67 $300 a month to lease
C-67 monthly service checks but no guaranteed repair service
Dyna DN-90 $200 a month to lease
Alman merged with Dup-Print; shows profits for last 3 years
X-112 back-to-back feature on any size paper
X-112 reduction 10, 20, and 30%
Dyna shows loss in last 2 years; *Business Week* notes cause as poor
    management
Conmo hailed by *Marketing Review* as promising newcomer in field;
    carries very small line but shows profit
C-76 back-to-back feature only on 8½- × 11-inch paper
C-76 reduction 10 and 20%
DN-90 call-in complaints 24-hour service guaranteed on any day
DN-90 back-to-back feature but 60-second adjustment time
DN-90 reduction 10% only
X-112 demonstration shows good quality copies on regular, back-
    to-back, and all reductions
DN-90 demonstration not available but sales representative brought
    quality copies to interview
C-76 demonstration shows good quality copies on all features but
    20% reduction

The notes may carry more detail than listed above, but the list represents major topics in our communication.

Once we have listed these items, the next step is to decide which items we want to discuss in relation to each other. One way of gathering related facts is simply to letter our list. The first item on the list we label "A"; if the next item relates to "A," we label it "A" also; if not, we label it "B." At this point we do not decide whether "A" will really come first in the communication. If the third topic relates to "A" or "B," we label it "A" or "B"; if not, we assign it the next letter, "C," and so on. If we letter our list, it would appear thus:

A  Alman X-112 $250 a month to lease
B  X-112 routine service every 3 months
B  X-112-24-hour repair service (guaranteed except for Fridays and weekends)
A  Conmo C-67 $300 a month to lease
B  C-67 monthly service checks but no guaranteed repair service
A  Dyna DN-90 $200 a month to lease
C  Alman merged with Dup-Print; shows profits for last 3 years
D  X-112 back-to-back feature on any size paper
D  X-112 reduction 10, 20, and 30%
C  Dyna shows loss in last 2 years; *Business Week* notes cause as poor management
C  Conmo hailed by *Marketing Review* as promising newcomer in field; carries very small line but shows profit
D  C-76 back-to-back feature only on 8½- × 11-inch paper
D  C-76 reduction 10 and 20%
B  DN-90 call-in complaints 24-hour service guaranteed on any day
D  DN-90 back-to-back feature but 60-second adjustment time
D  DN-90 reduction 10% only
E  X-112 demonstration shows good quality copies on regular, back-to-back, and all reductions
E  DN-90 demonstration not available but sales representative brought quality copies to interview
E  C-76 demonstration shows good quality copies on all features but 20% reduction

If we were working with many note cards, we might spread them out or stack them according to this method.

Now that we have labeled the items, we should gather all the related groups: the "A's," the "B's," and so on. Once we have gathered related facts, we need to decide how all the "A's" relate to each other, how all the "B's" relate, and so on. To do this, we can number or with a short list simply reorder, all the "A's," "B's," and so on. At the same time, we can decide what headings to assign the "A's," "B's," and so on. Finally, we should decide what overall order to give our outline. Will "A" really come first, or "C"? In the following list, we decided how the items in each lettered group relate, assigned each group a heading, and arranged the groups to reflect the final order of our outline:

Audience: Supervisor of Purchasing Department
Purpose: Investigate and recommend a duplicating machine to lease. Criteria: cost below $350 a month, back-to-back and reduction features. Method: interviews with sales representatives, demonstrations, annual reports and business journals.
Company Backgrounds (the C group)
    (C1) Alman merged with Dup-Print; shows profits for last 3 years.
    (C2) Conmo hailed by *Marketing Review* as promising newcomer in field; carries very small line but shows profit.
    (C3) Dyna shows loss in last 2 years; *Business Week* notes cause as poor management.
Cost to Lease (the A group)
    (A1) Alman X-112 $250 a month to lease.
    (A2) Conmo C-67 $300 a month to lease.
    (A3) Dyna DN-90 $200 a month to lease.
Required Features (the D group)
    Back-to-Back Copies
    (D1) X-112 back-to-back feature on any size paper.
    (D2) C-76 back-to-back feature only on 8½- × 11-inch paper.
    (D3) DN-90 back-to-back feature but 60-second adjustment time.
    Reduction Copies
    (D4) X-112 reduction 10, 20, and 30%.
    (D5) C-76 reduction 10 and 20%.
    (D6) DN-90 reduction 10% only.
Other Features and Services
    Quality of Copies (the E group)
    (E1) X-112 demonstration shows good quality copies on regular, back-to-back, and all reductions.
    (E2) C-76 demonstration shows good quality copies on all features but 20% reduction.
    (E3) DN-90 demonstration not available but sales representative brought quality copeis to interview.
Service (the B group)
    (B1) X-112 routine service every 3 months.
    X-112 24-hour repair service (guaranteed except for Fridays and weekends).
    (B2) C-67 monthly service checks but no guaranteed repair service.
    (B3) DN-90 call-in complaints; 24-hour service guaranteed on any day.
Conclusions and Recommendation of X-112

Since our outline evolved from our material, we can easily use it as a guide in writing our communication. Because we did not impose an outline upon our information but used our information to create one, the outline should reflect the logical order of our communication.

## Checking by a Formal Outline

Once we have created an outline from our information, we can use it to check the completeness, depth, order, length, and logic of our communication. Whether we use a topic outline (each item is a word or phrase) or a sentence outline (each item is the topic sentence for a paragraph or section in our communication), the advantages of placing our created outline into a formal structure are many. We have a final check to our approach; we can avoid overlaps in our discussion; we can spot gaps in our information; we can test the length and depth of our coverage; we can test the logic of our order.

The outline of the duplicating machine findings is logical, since it moves from background information necessary for this audience to essential criteria specified in our purpose to secondary criteria that might influence our decision to our final recommendation. In all cases, we have detailed information on the three choices with no overlaps, and when we have subdivided, we have at least two parts.

An outline, the final step before writing, evolves from information gathered from primary and secondary sources and is a final check on our research and thinking methods.

# Summary

After a thorough audience and purpose analysis, the next step is the prewriting process to gather and organize information. When using secondary sources to obtain information, we move from the general to the specific. Our research also helps us define and limit our topic. General guides lead us to select indexes and abstract journals which in turn direct us to specific sources. Even the card catalog can clue us to other sources. If our library offers computer-assisted searches, we can shorten our task once we have completed preliminary research. We transcribe our bibliography and notes onto note cards. We can quote directly, summarize, or write an objective or critical summary that we share with our audience.

We gather information from primary sources, such as interviews, questionnaires, and observations during experiments. Planning key questions, listening effectively, and summarizing during and after the interview ensure that we will gain the information we need. We can ask many people the same questions and compare their answers in a questionnaire. Most questionnaires consist of closed or multiple-choice questions. In relating information from an experiment, we discuss purpose, methods, observations, results, and conclusions.

The final step before writing is to create an outline. By a system of labeling and rearranging major topics or facts, we create an outline that evolves from our information, audience, and purpose. Placing this created outline in a formal structure enables us to check the completeness, depth, order, length, and logic of our communication. Now we are ready to write.

# Reading to Analyze and Discuss

## I

## On Being the Right Size

### J. B. S. Haldane

The most obvious differences between different animals are differences of size, but for some reason the zoologists have paid singularly little attention to them. In a large textbook of zoology before me I find no indication that the eagle is larger than the sparrow, or the hippopotamus bigger than the hare, though some grudging admissions are made in the case of the mouse and the whale. But yet it is easy to show that a hare could not be as large as a hippopotamus, or a whale as small as a herring. For every type of animal there is a most convenient size, and a large change in size inevitably carries with it a change of form.

Let us take the most obvious of possible cases, and consider a giant man sixty feet high—about the height of Giant Pope and Giant Pagan in the illustrated *Pilgrims Progress* of my childhood. These monsters were not only ten times as high as Christian, but ten times as wide and ten times as thick, so that their total weight was a thousand times his, or about eighty to ninety tons. Unfortunately the cross sections of their bones were only a hundred times those of Christian, so that every square inch of giant bone had to support ten times the weight borne by a square inch of human bone. As the human thigh-bone breaks under about ten times the human weight, Pope and Pagan would have broken their thighs every time they took a step. This was doubtless why they were sitting down in the picture I remember. But it lessens one's respect for Christian and Jack the Giant Killer.

To turn to zoology, suppose that a gazelle, a graceful little creature with long thin legs, is to become large, it will break its bones unless it does one of two things. It may make its legs short and thick, like the rhinoceros, so that every pound of weight has still about the same area of bone to support it. Or it can compress its body and stretch out its legs obliquely to gain stability, like the giraffe. I mention these two beasts because they happen to belong to the same order as the gazelle, and both are quite successful mechanically, being remarkably fast runners.

Gravity, a mere nuisance to Christian, was a terror to Pope, Pagan, and Despair. To the mouse and any smaller animals it presents practically no dangers. You can drop a mouse down a thousand-yard mine shaft; and, on arriving at the bottom, it gets a slight shock and walks away, provided that the ground is fairly soft. A rat is killed, a man is broken, a horse splashes. For the resistance presented to movement by the air is proportional to the surface of the moving object. Divide an animal's length, breadth, and height each by ten; its weight is reduced to a thousandth, but its surface only to a hundredth. So the resistance to falling in the case of the small animal is relatively ten times greater than the driving force.

An insect, therefore, is not afraid of gravity; it can fall without danger, and can cling to the ceiling with remarkably little trouble. It can go in for elegant and fantastic forms of support like that of the daddylonglegs. But there is a force which is as formidable to an insect as gravitation to a mammal. This is surface tension. A man coming out of a bath carries with him a film of water of about one-fiftieth of an inch in thickness. This weighs

**37**

roughly a pound. A wet mouse has to carry about its own weight of water. A wet fly has to lift many times its own weight and, as everyone knows, a fly once wetted by water or any other liquid is in a very serious position indeed. An insect going for a drink is in as great danger as a man leaning out over a precipice in search of food. If it once falls into the grip of the surface tension of the water—that is to say, gets wet—it is likely to remain so until it drowns. A few insects, such as water-beetles, contrive to be unwettable; the majority keep well away from their drink by means of a long proboscis.

Of course tall land animals have other difficulties. They have to pump their blood to greater heights than a man, and, therefore, require a larger blood pressure and tougher blood-vessels. A great many men die from burst arteries, especially in the brain, and this danger is presumably still greater for an elephant or a giraffe. But animals of all kinds find difficulties in size for the following reason. A typical small animal, say a microscopic worm or rotifer, has a smooth skin through which all the oxygen it requires can soak in, a straight gut with sufficient surface to absorb its food, and a single kidney. Increase its dimensions tenfold in every direction, and its weight is increased a thousand times, so that if it is to use its muscles as efficiently as its miniature counterpart, it will need a thousand times as much food and oxygen per day and will excrete a thousand times as much of waste products.

Now if its shape is unaltered its surface will increase only a hundredfold, and ten times as much oxygen must enter per minute through each square millimetre of skin, ten times as much food through each square millimetre of intestine. When a limit is reached to their absorptive powers their surface has to be increased by some special device. For example, a part of the skin may be drawn out into tufts to make gills or pushed in to make lungs, thus increasing the oxygen-absorbing surface in proportion to the animal's bulk. A man, for example, has a hundred square yards of lung. Similarly, the gut, instead of being smooth and straight, becomes coiled and develops a velvety surface, and other organs increase in complication. The higher animals are not larger than the lower because they are more complicated. They are more complicated because they are larger. Just the same is true of plants. The simplest plants, such as the green algae growing in stagnant water or on the bark of trees, are mere round cells. The higher plants increase their surface by putting out leaves and roots. Comparative anatomy is largely the story of the struggle to increase surface in proportion to volume.

Some of the methods of increasing the surface are useful up to a point, but not capable of a very wide adaptation. For example, while vertebrates carry the oxygen from the gills or lungs all over the body in the blood, insects take air directly to every part of their body by tiny blind tubes called tracheae which open to the surface at many different points. Now, although by their breathing movements they can renew the air in the outer part of the tracheal system, the oxygen has to penetrate the finer branches by means of diffusion. Gases can diffuse easily through very small distances, not many times larger than the average length travelled by a gas molecule between collisions with other molecules. But when such vast journeys—from the point of view of a molecule—as a quarter of an inch have to be made, the process becomes slow. So the portions of an insect's body more than a quarter of an inch from the air would always be short of oxygen. In consequence hardly any insects are much more than half an inch thick. Land crabs are built on the same general plan as insects, but are much clumsier. Yet like ourselves they carry oxygen around in their blood, and are therefore able to grow far larger than any insects. If the insects had hit on a plan for driving air through their tissues instead of letting it soak in, they might well have become as large as lobsters, though other considerations would have prevented them from becoming as large as man. . . .

# Questions for Discussion

1. What is the thesis or main idea of Haldane's essay?

2. Who is Haldane's audience? What evidence do you find for your decision?

3. What methods does Haldane use to illustrate his thesis? How familiar are his examples?

4. Why is gravity a problem for larger animals and not for smaller animals? What force is a problem for smaller animals and not for larger ones? What accounts for these differences?

5. What are the disadvantages of height? What methods are used to increase surface area in large animals and plants? Why are some of these methods not adaptable by smaller animals?

# Exercises and Assignments

1. **Problem:** Reread the Haldane essay, "On Being the Right Size." Underline or record on note cards the main ideas of the essay. Mark significant quotations.

Assignment:

1. Write an objective summary of the essay. Choose an appropriate audience and convey to that audience the content, style, structure, and tone of the essay.
2. Write a critical summary of the essay. Choose an appropriate audience and not only convey the essence of the essay but also *evaluate* the structure, style, and content.
3. Outline the essay in a formal topic outline. Write a one-paragraph evaluation of the logic and coverage of the essay as represented by the outline.

2. **Problem:** Research in secondary sources helps us narrow our project topic as we turn to more and more specific sources. This exercise will take you from a broad topic and a general source to a narrow topic and a specific secondary source. Choose a topic within your major or a technical or business topic you are interested in. The topic can be somewhat broad, such as ecology, computers, or income tax.

Assignment

1. Find your broad topic in a general guide to secondary sources such as Sheehy or Coman. List five major secondary sources that cover your topic. In some cases, you may have to find your topic included in a broad field, for example, ecology under biology or environmental biology.
2. Look up your topic in one of the major secondary sources you found in step 1. The source will probably be an index or abstract journal. At this point, consider how you can narrow your topic. The range of books and articles you find in the index or abstract journal will help you. For example, you might redefine ecology as air pollution controls. List five secondary sources you find listed in the index or abstract journal that covers your narrow topic. Some of the five should be books and some periodicals.
3. From your list of five secondary sources, locate in the library one periodical article. At this point narrow your topic so you could discuss it in a hypothetical five-page paper, for example, the success of new air pollution controls for cars. Take notes on note cards on the content of the periodical article.

3. **Problem:** Once we have completed research in secondary sources and have defined and narrowed our project topic, we can do original research.

## Assignment

**1.** Using the same topic as in Problem 2, set up a hypothetical interview with someone who could give you new information; for example, for research on the success of air pollution controls for cars, you might want to interview the head of the Environmental Protection Agency. List the purpose of the interview and your key questions.

**2.** Using the same topic as in Problem 2, write a questionnaire that you can send to a large audience; for example, for your research on the success of air pollution controls for cars, you might want responses from car customers or dealers. Use as many closed questions as possible.

**4. Problem:** You have been asked by your supervisor to investigate the possibility of setting up an advanced training program for office and sales representatives. The program can either be an in-house program, or the company can take advantage of the courses offered at the local community college. Your supervisor has given you some ideas, and you have interviewed employees and department heads to assess their needs. You have read the college catalog and interviewed some community college instructors. You have taken notes during the interviews and have notes from the catalog and are now ready to write your report.

## Assignment:

**Create an outline for your report to your supervisor from the following random notes. Include in your outline your recommendations.**

Most sales managers have college degrees in business and accounting.

Sales representatives often must see clients at night.

Office employees: secretaries (32) and clerks (20).

Some secretaries want to take college courses in business administration.

Office employees all have high school diplomas.

75 sales managers.

150 sales representatives.

Secretaries are willing to take night courses.

Clerks feel no need for advanced training unless a promotion is guaranteed after training.

Secretaries need advanced training in shorthand.

Clerks are willing to take night courses.

Supervisors state that 5% of clerks are capable of moving into secretarial positions after more training in "secretarial science."

Most community college courses offered for three credit hours.

52 office employees.

Sales representatives have a high school degree at least.

Sales representatives want courses in advertising copy writing.
Community college courses cost $15 per credit hour.
Few sales representatives have completed college.
Sales representatives want advice from professionals in selling techniques.
Community college offers complete program in secretarial science.
Some secretaries want more education in business administration.
Most sales managers have had experience in selling techniques before being promoted to management positions.
Community college offers courses in advertising copy writing and in advanced shorthand.
Community college offers a course in introduction to business administration.

# PART TWO

## Identifying a Subject for an Audience

# 3

# **Definition:**
## Distinguishing a Subject from Others

*Defining a subject informally.*
*Defining a subject formally.*
*Writing a stipulatory or operational definition.*
*Amplifying or extending a definition.*
*Defining a subject for various audiences and purposes.*
*Reading and analyzing definitions written by others.*

In technical communication, we rely on definition to help our audience understand our subjects and in what special sense we may use them. Definition clarifies our subject matter and our approach. Because definition establishes a common vocabulary for the audience and the writer, it serves as a natural introduction to a communication.

Definition distinguishes or differentiates a subject from all other similar ones. It states what characteristics the subject possesses that no other subject, even in the same class, possesses.

For example, in the early part of the twentieth century, Sigmund Freud opened a lecture on psychoanalysis with a basic definition of the subject. Psychoanalysis is a medical treatment, but it differs from all others because it is used to treat "those suffering from nervous disorders":

> One thing, at least, I may presuppose that you know—namely, that psychoanalysis is a method of medical treatment for those suffering from nervous disorders; and I can give you at once an illustration of the way in which psychoanalytic procedure differs from, and often even reverses, what is customary in other branches of medicine. Usually, when we introduce a patient to a new form of treatment, we minimize its difficulties and give him confident assurances of its success. This is, in my opinion, perfectly justifiable, for we thereby increase the probability of success. But when we undertake to treat a neurotic psychoanalytically we proceed otherwise. We explain to him the difficulties of the method, its long duration, the trials and sacrifices which will be required of him; and, as to the result, we tell him that we can make no definite promises, that success depends upon his endeavors, upon his understanding, his adaptability, and his perseverance. . . .

Freud defined psychoanalysis for his peers by distinguishing it from other forms of medical treatment. In technical communication, we use definition to establish a common vocabulary, clarify the meaning of our subject, and distinguish it from similar subjects. We usually define our subject in the beginning of a communication.

## Informal Definition

When we define a subject informally, we convey the meaning quickly and do not stress the definition. We might decide that our audience is familiar with the subject and needs only a reminder of the definition, or we might have defined the subject earlier and just assure the audience that we are using the subject in the same sense. Informal definitions are often synonyms that are more familiar to the audience than the subject being defined.

For example, the following terms are defined informally by synonyms or brief descriptions containing synonyms (italics):

Bacteria, *germs,* may be harmful or beneficial.
Heterogeneous groups, *groups with different ideas or characteristics,* are more likely to experience hostility, than are homogeneous, or *similar,* groups.

One advantage of the informal definition is that it reads as a smooth, integral part of the text.

***An Example of Informal Definition.*** In the following introduction to a memo on processing medical claims for obesity, the writer defines informally exogenous and endogenous. Although he reminds his audience what these terms mean, he centers his discussion on the difficulty in defining his main subject, obesity:

> This memorandum provides some guidelines for the administration of claims for obesity, whether *exogenous,* resulting from an outside factor, such as overeating, or *endogenous,* due to inside factors such as metabolic or endocrinic disorders.
>
> It must be clearly understood at the outset of this memorandum that determining when obesity can be classified as a bodily disorder and, therefore, an "illness" under our standard definition of that term does not easily lend itself to codification. Medical opinion continues to be divided over whether obesity of and by itself does, in fact, constitute a clinically recognizable "illness.". . .

The writer defines exogenous and endogenous without interrupting his discussion of obesity and his search for a practical definition of that term. Thus we define subjects informally when our audience is somewhat familiar with them or when we do not need to stress our definition. However, if we introduce a new subject or define a subject in a new way, we should define it formally.

# Formal Definition

Formal definitions place the subject being defined within a class or genus and then distinguish that subject from all others within the same class. What distinguish the subject being defined are the *differentiae.* Thus we place our subject in a class of subjects the audience may be familiar with and then tell what makes our subject different from all the others in that class. Formal definitions remind our audience of what they know about our subject by looking at similar ones, and then explain what is unique or important about our subject by distinguishing it.

We define *formally* new or unfamiliar subjects for our audiences because the formal definition is a word equation, precise and complete.

The term to be defined = the class + the differentiae.

Again a formal definition consists of these parts:

1. The subject to be defined
2. The class, genus category, or group to which the subject belongs
3. The differentiae or the characteristics that distinguish the subject from all others within the same class and exclude all other subjects but the one being defined

For example, the following is a formal definition:

| man | is | a rational | animal |
|-----|-----|------------|--------|
| | | (differentiae) | (class) |

If man belongs to the genus or class "animal" and is indeed the only one of that class who is "rational," then the parts of the formal definition are equal and complete. The following are formal definitions:

| Subjects | Class | Differentiae |
|----------|-------|--------------|
| school | is an institution | for instruction |
| biology | is the study or science | of living organisms and vital processes |
| science | is the investigation and analysis | of natural phenomena |
| mass medium | is a communication channel | designed to reach large audiences, to transmit an impersonal message, and to stimulate indirect feedback |

## Rules of Formal Definition

The test of a formal definition is whether (1) the subject does belong to the class and (2) the differentiae exclude all members of that class except the one being defined. For example, the differentiae for a school must exclude all other institutions such as hospitals, churches, and marriages. Science must be distinguished from technology, the application of science, and biology must be distinguished from the other sciences, chemistry and physics. Obviously the most difficult part of the formal definition to convey to an audience is the differentiae.

The terms we use in our differentiae should be familiar to our audience, or we should define these terms. We might also identify the components or origins of our subject. For example, before we define a "mass medium," we might explain that a medium is "an intervening substance or agent through which something is transmitted," in this case a message. Moreover, our differentiae should not repeat a word used in the class or the subject being defined, or our definition does not provide information. For example, the definition "hockey is a game played on a hockey rink" does not define but repeats. Following these guidelines, using familiar terms in the differentiae, explaining the origins of the subject, and avoiding repetition, we can be sure that our formal definitions introduce clearly and completely our subjects to our audiences.

## Characteristics of Differentiae

Differentiae can include four characteristics or "causes": the *efficient* cause (what created the object or term), the *material* cause (what the object is made of), the *formal* cause (what the structure of the object is), and the *final* cause (what the function of the object is). For example, a desk (*subject* to be defined) is a piece of furniture (*class*) made by a carpenter (*efficient* cause) from wood or metal (*material* cause), composed of a flat top placed horizontally upon four legs (*formal* cause), and used for writing, typing, or studying (*final* cause). However, differentiae need not include all of these causes. Our choice depends on our subject and on our audience's needs.

Again we define formally new and unfamiliar subjects or major subjects in a communication. We affiliate our subject with a class to emphasize similarities with more familiar subjects, and then identify the unique characteristics of our subject by differentiae. In essence our audience first sees what is "old" and then what is "new" about our subject.

**Examples of Formal Definition.** We can see this technique in contemporary technical communication. Because the writers have used formal definitions, their audiences know the exact meanings of the terms used in the communications.

from a research report on "Multi-Species Fish Farming":

> Polyculture, the rearing of combinations of two or more non-competitive fish species in confinement, has considerable economic advantage over raising a single species for food production.

from a technical paper on "Low Stress Grinding":

> Low stress grinding is an abrasive material process that leaves a low magnitude, generally compressive, residual stress in the surface of the workpiece.

from a catalog offering control and value products:

> The Type 2340-390 is a 2-wire transmitter that senses gauge pressure with a Bourdon tube. it produces a standard 4 to 20 mA or 10 to 50 mA signal with input spans as small as 3 psi or as large as 20,000 psi.

Thus a formal definition places a subject within a class and then distinguishes that subject from all others. We use formal definitions to convey the exact meaning of a subject before we begin a discussion of that subject.

# Stipulatory Definition

Sometimes we define a familiar subject in a broader or narrower sense than usual, or we confine the subject to a special time or place. We need to redefine our subject by a "stipulatory," or "operational," definition. If we announce our stipulations initially, our audience will usually accept our special definition. In this sense our stipulatory definition promises that we will cover so much of a subject, no more and no less; it announces a specific approach to the subject. We often use stipulatory definitions in the beginning of a technical report; we define the problem we investigated, the new device we developed, the technical terms we use, and so on, in a special sense.

These limitations we place on our subjects can be time and space qualifications. For example, we might explore progress in a certain year, or within a certain geographical location, and we would announce these limitations in a stipulatory definition. We might investigate compact cars produced in the United States in the years 1965–1975. We would define "compact cars" as they appeared in that time and place. Stipulatory definitions also can narrow or broaden a subject. For example, we might confine a report on schools to elementary schools, and our definition would be: "Within this report, a school is considered an institution for the instruction of *young* people between the ages of 6 and 12." Or we might consider self-generating schools or study groups: "Within this report, a school is considered an institution for the instruction of people with a similar philosophy and purpose." Notice that the class usually does not change; the differentiae contain the qualifications or self-imposed limitations.

Any definition, whether stipulatory or not, should still be objective. While audiences will accept our self-imposed limitations, they will not accept bias or personal prejudice. We should not insist on debatable meanings; for example, the definition "Science is the control and manipulation of natural phenomena" reflects a negative and biased attitude toward science.

**An Example of Stipulatory Definition.** While one can define "proper management" in many ways, in the following example, the writers stipulate that proper management pertains to a proper environment for laboratory animals:

> The proper management of laboratory animal facilities depends on many subjective and objective factors that interact differently in different institutions. Well-trained and motivated personnel can ensure high-quality animal care even in the presence of deficiencies in the physical plant or housing equipment. For the purposes of the *Guide,* "proper management" is defined as any system of housing and care that permits animals to grow, mature, reproduce, and behave normally and to be maintained in physical comfort and good health. "Proper management" also implies environmental and genetic control to minimize variations that may modify an animal's response to a particular experimental regimen. Proper management of laboratory animals is essential to the welfare of the animals, to the validity of research data, and to the health and safety of the animal-care staff.

The stipulatory definition above gives a special meaning for a common term and serves as a natural introduction to the guide on laboratory animal care for peer professionals. Thus in stipulatory definition, we place time or space limitations on our subject or define our subject in a broader or narrower sense than usual. Stipulatory definitions, like standard formal definitions, are a natural introduction to a communication.

# Amplified or Extended Definition

While definition serves as an introduction to any discussion, it can also form the body of a communication. We amplify or extend formal definitions when we need to discuss fully a new subject, contrast it to an old definition, give examples of how our definition applies, and so on. We support definition then by many of the rhetorical patterns we will study in depth in this text. However, now let us look at some of the ways we can amplify definition when our main *purpose* is to *define.*

## Ways of Extending Definition

We can amplify or extend definitions by the following means:

1. Further definition: The subjects implied or stated within the definition defined. For example, the definition of mass media might include a definition of media.
2. Genesis or origin of the subject: The derivation and historical meaning of the subject. For example, *media* is Latin for the middle (a sense of intermediate).
3. Examples: Representative examples of the subject and how it is used. For example, mass media includes newspapers, radio, and TV.
4. Comparison-contrast: Comparing the subject with and contrasting it to another. For example, mass media might be contrasted to interpersonal communication.
5. Structural analysis: A division and examination of the subject or concept and its components. For example, mass media consists of the source, the message, the channel, the receiver, and the feedback.
6. Causal analysis: The cause and effects of a subject. For example, a message in the mass media, such as violence, can affect an audience in certain ways.
7. Description: An examination of the features of a subject at rest or in motion. For example, a message is created and sent within mass media.
8. Principles: By what concept or universal principle or law a subject might operate. For example, the principle of source-receiver explains mass media.
9. Classification: The groups or categories of a subject. For example, mass media can be divided into the print media and the electronic media.

We can use these and other means alone or in combination to amplify a definition. An extended definition serves as a thorough introduction to or discussion of any subject. Let us look at some examples of extended definition.

## Examples of Extended Definition

*Extension by Further Definition*   The following definition of chemistry could stand alone as a formal introduction to the term:

> Chemistry is a branch of science which deals with the study of matter, or in other words with the character of the "stuff" of which the material universe is composed.

However, the writer expanded this definition to include the past and present tasks of the chemist and the definition of chemical theory (defined formally) and chemical laws and hypotheses (defined informally):

> It is obvious to the senses that matter abounds around us in many kinds of forms. It is the task of chemistry to separate from this heterogeneous assemblage of matter, various homogeneous portions known as substances, each of which has its individual composition and properties. A vast mass of information of this kind has been accumulated as a result of patient experiments and observations. The observed characters and inter-relationships have led to many consequences, notably to the classification of substances, to the preparation of one known substance from another, and to the elaboration of new substances unknown in nature. Prolonged work of this kind was necessary before any accurate idea could be formed of the proximate or ultimate nature of matter. In the modern development of chemistry facts, accurately established by experiment, led to the formulation of generalized statements, or *laws*. Certain laws, fitted by imaginative processes of thought into a wider conception, gave birth to *hypotheses*, which when fully established took higher rank as *theories*. Theories are thus conclusions drawn from accumulated facts and capable of leading to the prediction of new facts.

The writer's main purpose in the passage above is to derive a definition of chemistry that reflects modern development of chemical facts. Often we extend or amplify definitions to introduce a new meaning or use of our subject.

*Extention by Examining Origins.*   Tracing the origins of a subject also clarifies a definition for an audience. In the following passage, Isaac Asimov explains the derivation of the word "ether" from the Greek word for heavenly material:

> To Aristotle the manner in which an object moved was dictated by its own nature. Earthy materials fell and fiery particles rose because earthy materials had an innate tendency to fall and fiery particles an innate tendency to rise. Therefore, since the objects in the heavens seemed to move in a fashion characteristic of themselves (they moved circularly, round and round, instead of vertically, up or down), they had to be made of a substance completely different from any with which we are acquainted down here.
>
> It was impossible to reach the heavens and study this mysterious substance, but it could at least be given a name. (The Greeks were good at making up names, whence the phrase, "The Greeks had a word for it.") The one property of the heavenly objects that could be perceived, aside from their peculiar motion, was, however, their blaz-

ing luminosity. The sun, moon, planets, stars, comets, and meteors all gave off light. The Greek word for "to blaze" (transliterated into our alphabet) is *aithein.* Aristotle therefore called the heavenly material *aither,* signifying "that which blazes." In Aristotle's day it was pronounced "i'ther," with a long *i.* . . .

This Aristotelian sense of the word *ether* is still with us whenever we speak of something that is heavenly, impalpable, refined of all crass material attributes, incredibly delicate, and so on and so on, as being "ethereal."

Asimov traced the origins of ether to stress the broad meaning of the term; it can mean much more than an organic compound. Understanding the origins of a subject helps our audience appreciate how we use the subject now.

**Extension by Principle and Description.**  Describing the subject in use and explaining the principles by which it works also make a subject concrete for an audience. In the following definition, ultrasonic grinding and the "chipping action" by which the process works are described:

Ultrasonic Grinding (Impact Grinding or Ultrasonic Machining USM) is a process by which material is removed from the workpiece with the use of a high frequency tool which has the same shape of the material to be removed. Such a tool is pressed onto the workpiece by a preset light pressure and then applying a high frequency low amplitude vibration. At the same time, an abrasive suspension is continually fed at the tool-workpiece interface. The wearing or the cutting of the workpiece can be explained as a result of a "chipping action." The abrasive grains are constantly being pushed or crushed against the workpiece by the oscillating tool. Thus, the abrasive grains cause the workpiece to erode into the desired shape. Due to the nature of Ultrasonic Grinding, not only irregular and intricate shapes are obtainable, but materials harder than $R_c$ 60 can be machined. Also USM unlike Electric Discharge Machining, for example, can be used to machine both metallic and nonmetallic materials.

The audience understands exactly what ultrasonic grinding is before seeing how it works. The writer defines the process so that his peer professional audience can anticipate and eventually duplicate the process.

We extend or amplify a definition to clarify our subject. We can give further definitions, trace the origins of our subject, give concrete examples, describe the subject at rest or in motion, explain the principles by which it works, compare and contrast our definition with a past one, or analyze and classify the causes and categories of our subject.

# Audiences and Purposes of Definition

As we have seen in previous examples, definition serves as natural introduction to technical communications. The extent of our definitions and the means by which we amplify them are determined by our audience and purpose. We have seen a variety of definitions for technical and general audiences. Let us look more closely at the audiences and purposes of definition.

## Technical Audiences

As we know, when addressing a technical audience, we can use a special vocabulary and give theoretical explanations. The following definition introduces guidelines on the safe use of the chemical methylamine. Although the audience is a technical one, the writers still use a formal definition amplified by comparing and contrasting methylamine to ammonia:

> The methylamines are highly reactive, water-soluble gases which are used primarily as intermediates for the production of a wide variety of products. These derivatives of ammonia and methanol, which are the most economical source of organic amine for synthesis, are available as anhydrous liquefied gases in pressurized containers or as aqueous solutions of various concentrations.
>
> The physical properties of the methylamines are similar to those of ammonia. Like it, they are toxic; but unlike it, they are highly flammable. Also, very low concentrations of the methylamines in the air will produce an unpleasant odor not unlike that of dead fish. Therefore when using these compounds, the protection of health, property, and the quality of the environment depend on a thorough understanding of their properties and of proper handling procedures.

Specifications of anhydrous methylamine followed this definition. Although the writers use a specialized vocabulary, they extend and clarify their definition by comparison-contrast. The definition helps the audience decide whether to use these chemicals; the audience understands exactly what methylamines are, their use, and the safety factors involved.

## Nontechnical Audiences

On the other hand, when we define a subject for a general or lay audience, we use more concrete examples and a more common vocabulary. The purpose of the following definition is to make the public more aware of and sensitive toward a rare affliction. Although the student writer gives the formal medical definition of acromegaly, she clarifies the subject by vivid examples and contrasts a person's state before and after affliction. The formal definition establishes a common vocabulary and clarifies the term; the extension of the definition stresses the effects of acromegaly on daily life:

> Acromegaly—A glandular disorder characterized by an enlargement of the extremities, thorax, and face, including both soft and bony parts. Such is the dictionary definition of this affliction. If only it were that simple, but the definition doesn't even begin to cover this rare irregularity which turns a man into a giant—literally.
>
> A dwarf can sometimes be given hormones to stimulate growth, but a giant cannot be shrunk. Once the growth process has taken

place it is irreversible. The damage has been done. The afflicted now dwarfs his fellowmen. He's a freak in nature and people can't help staring, nor can doctors resist the opportunity to look, poke, and prod.

The result is obvious. What is the cause? The medical world knows little about this phenomenon of growth. In fact, doctors have been known to tell the afflicted that "you don't grow fat on thin air." (How unwittingly right they are.) The implication is that the individual is overeating and must diet; meanwhile it is not the individual's fault, but his gland's. Possibly there's a cyst on the gland, or a tumor pressing upon it. Maybe it's just the gland itself, deficient in its normal function. Tests are sometimes conclusive, and at other times inconclusive. Whatever has happened, the result is that the gland has gone "berserk" sending out growth hormones as quickly as it can get them into the system.

A shoe size goes from an eight to a thirteen in a relatively short time of a year or two. Hands grow bigger, arms lengthen, and the body shoots up past the six-foot mark. Handsome features become distorted with enlargement. The physical stress of acromegaly is hard enough, but the mental strain is brutal. How does a "David" suddenly cope with turning into a "Goliath"?

*MaryBeth Dennett*

The writer uses a vivid and provocative yet nontechnical vocabulary to define this technical subject for a general audience. She offers concrete and familiar images to help the audience picture the effects of the disease.

# Summary

Definition clarifies for our audience our subject and our approach to it. Definition distinguishes by contrast a subject from all similar subjects in the same class, and serves as a natural introduction to most technical communication.

We can define a subject informally by synonyms or formally by stating the class and differentiae of the subject. Because the differentiae must exclude all other subjects but the one being defined, it is the most difficult part of a definition to write. In essence the class tells the audience what is familiar about the subject and the differentiae what is unique. Stipulatory definition qualifies a subject by time and place, narrow or broad approach.

We can amplify a definition to clarify our subject by many of the rhetorical patterns we will study in depth in later chapters. The extent to which we amplify and the means by which we do so are determined by our audience and purpose.

# Reading
## to Analyze and Discuss

# I

## A

For the purpose of this report, a medical-information system is computer-based and receives data normally recorded about patients, creates and maintains from these data a computerized medical record for every patient, and makes the data available for the following uses: patient care, administrative and business management, monitoring and evaluating medical-care services, epidemiological and clinical research, and planning of medical-care resources.

## B

While machinability has been defined many different ways, the definition usually used in comparing the machinability of different materials is

$$I = \frac{V_{t20} \text{ of the material being tested}}{V_{t20} \text{ of the standard material (AISI B1112)}} \times 100$$

where I = machinability rating or index

$V_{t20}$ = cutting velocity which provides a 20-minute tool life

In most cases tool life is defined as the length of time the tool cut before a given-size wearland occurs on the cutting tool. . . .

## C

Gilbert N. Lewis once defined physical chemistry as "anything that's interesting." Although this may seem a little unfair to the rest of chemistry, more specific definitions are hard to come by. Physical chemistry is the foundation for every other branch of chemistry, and it merges so gradually into the various superstructures that it's almost impossible to make a clean distinction between physical chemistry and bio-chemistry, say, or polymer chemistry.

Perhaps as good a definition as any, and one with particular historical significance, is that physical chemistry is the study of the physical principles underlying chemical phenomena. More and more over the past several decades, chemists have taken to exploring the properties of matter from the molecular and electronic viewpoint. In the process, they have eagerly appropriated instruments and techniques from physics. Nuclear magnetic resonance spectrometry is one obvious example. More recent entries include picosecond lasers and molecular beams.

As a result, the distinction between physical chemists and chemical physicists has all but vanished. About one half the division chairmen in the American Physical Society also have been division chairmen in the

American Chemical Society. About 60% of the authors in the *Journal of Chemical Physics* work in chemistry departments. The only real distinction, jokes one chemist, is that to a physicist, all molecules are the same: simple manifestations of the Schroedinger equation. But a chemist appreciates the difference.

## Questions for Discussion

1. Identify the types of definitions (informal, formal, and stipulatory) that appear in the three selections. What methods have been used to extend these definitions?

2. What are the audiences of the three selections? What is the purpose of the communications they appear in?

3. In selection A, could another writer define a "medical information system" in another way? If so, how?

4. In selection B, identify the class and differentiae. Do the differentiae exclude all other members of the same class?

5. In selection C, why is it difficult for the writer to define physical chemistry? What differentiates this science from all others?

# Exercises and Assignments

**1. Problem:** When defining a subject, first we place it in a class with which our audience may be familiar. Knowing the proper class for our subject helps our audience begin to contrast our subject with all others in the same class.

Assignment:
Place the following terms in their proper class.

| Subject | Class |
|---|---|
| inch | _____ |
| automobile | _____ |
| chemistry | _____ |
| solid | _____ |
| spider | _____ |
| molecule | _____ |
| microscope | _____ |

**2. Problem:** The most difficult part of a definition to write is the differentiae. The differentiae must distinguish completely the subject from all others in the same class.

Assignment:

**1.** Name the class in which each of the following pairs of subjects belong.
**2.** List the differences between the subjects in each pair. These differences would form the differentiae for each subject.

biology—physics
salt—sugar
bicycle—motorcycle
collie—beagle
optician—ophthalmologist
poker—bridge
football—basketball
chemical engineering—mechanical engineering

**3. Assignment:**

Supply the differentiae for the following definitions.

| Subject | Class | Differentiae |
|---|---|---|
| pulse | a regular throbbing or beat caused by | _____ |
| telegraph | an apparatus of communication that | _____ |
| economics | a social science concerned with | _____ |
| electron | an elementary particle consisting of | _____ |
| foreign affairs | matters having to do with | _____ |

**4. Problem:** We may use different terminology when defining a subject for a layperson from that used when defining the same subject for a technologist or scientist.

**Assignment:**

**1.** Define the following subjects *formally first* for the technologist or scientist and *then* for the layperson.

| Subject |
| --- |
| nucleus |
| tuberculosis |
| energy |
| photosynthesis |
| computer |
| liquid |

**2.** Assuming that you have already defined formally the subjects above in a communication for a technologist, write an informal definition of each that might be included later in the communication to remind the audience of the meaning of the subject.

**5. Problem:** We can extend or amplify a definition of a subject by many rhetorical patterns. An amplified definition may serve as the introduction to or the body of a communication.

**Assignment:**

**1.** Review the ways of extending a definition.
**2.** Write a formal extended definition of a subject of your choice.
**3.** Describe the audience and purpose of your communication.

**6. Problem:** A stipulatory definition qualifies a subject by time and place, or identifies a narrower or broader approach than usual.

**Assignment:**

**1.** Write a stipulatory definition to serve as an introduction to a technical communication.
**2.** Describe briefly the content, purpose, and audience of the communication.

# 4

# Description:
## Helping an Audience
## Visualize a Subject

*Describing the overall appearance of a device.*
*Describing the parts of a device.*
*Describing a device for various audiences and purposes.*
*Achieving precise wording in device description.*
*Supporting device description with graphic aids.*
*Reading and analyzing device descriptions written by others.*

One basic task of technical communicators is to describe a subject, whether an object, a machine, an organism, or an organization, so that the audience can visualize it clearly and accurately. When we first see a person, our initial impression is often based on appearance. Similarly, before we can describe how a mechanism works, before we can tell how to operate or repair a device, before we can contrast subjects or analyze their causes and effects, we need to give our audience the sense of orientation that comes from a detailed description. To do this we freeze the motion or action of our subject. We isolate our subject and concentrate our attention on the subject and its "parts."

Perhaps no other rhetorical pattern demands as much precision in language as does description. We need to convey to our audience the overall appearance of our subject, the dimensions of its parts, the materials that compose them, and how and where the parts are attached. Our words must draw an accurate picture of our subject; technical description relies on precise figures and words. In this chapter we concentrate on device description, the most common yet challenging form of technical description.

The first communications of early scientists and technologists were often based on descriptions of their subjects. René Laënnec's description of the stethoscope he invented in the early part of the nineteenth century is a classic example of device description. He describes not only the size and appearance of the stethoscope but also its function:

> I consequently employ at the present time a wooden cylinder with a tube three lines in diameter bored right down its axis; it is divisible into two parts by means of a screw and is thus more portable. One of the parts is hollowed out at its end into a wide funnel-shaped depression one and one half inches deep leading into the central tube. A cylinder made like this is the instrument most suitable for exploring breath sounds and râles. It is converted into a tube of uniform diameter with thick walls all the way, for exploring the voice and the heartbeats, by introducing into the funnel or bell a kind of stopper made of the same wood, fitting it quite closely; this is made fast by means of a small brass tube running through it, entering a certain distance into the tubular space running through the length of the cylinder. . . .

Upon reading Laënnec's precise device description anyone can duplicate his stethoscope. The test of an accurate description is if the audience can reproduce the device either literally, by making it, or visually, by drawing the device after reading the description.

# The Pattern of Description

Description moves from the general to the specific, from the whole to the parts. At each stage in this movement, we mention characteristics such as size, shape, material, location, and method of attachment. Description demands precision in language to convey the appearance of the subject and its parts. A description then helps an audience visualize a subject by drawing *in words* an accurate picture of it.

Notice the exact detail, especially in dimensions and in location and attachment of parts, of the following description of a Vanguard magnetic field satellite. The description moves from the satellite as a whole to the specific parts of the satellite:

> The Magnetic Field Satellite was a 13-inch-diameter fiberglass sphere with a 2⁷⁄₁₆-inch-diameter fiberglass cylinder protruding 17¼ inches from the top, and with a 2¾-in-diameter magnesium cylinder protruding 1⅝ inches from the bottom. The total length was 31¾ inches. Four antennas were equally spaced around the satellite's equator. Each was ⅜ inch in diameter at the base, ¼ inch at the tip, and 23½ inches long.
>
> Inside the sphere was a cylindrical magnesium container with two pressure-tight compartments. One contained batteries to power the satellite electronic equipment, while the "upper" compartment housed the magnetometer electronics module and a second module containing the transmitters and the command receiver. The magnetometer sensing unit was positioned at the outer end of the fiberglass tube. . . .

After describing the general appearance of the satellite, the writer looked at the parts inside the satellite. In description we move from general to specific parts in a definite direction: from outside to inside, top to bottom, left to right, and so on.

## Characteristics of Device Description

Device description usually follows a specific pattern of development while tracing the characteristics of the whole subject and its parts. This pattern reflects the minimum requirements of device description; the final arrangement and amount of detail depend on the audience and purpose of the description. In the introduction to a device description, we usually define the subject, state its function, describe the overall size and shape, and identify the subject.

*Introduction.* We begin description with a definition to distinguish our subject from all others in the same class. Definition is a natural opening for description because definition moves from the general or the class to the specific or the differentia. Then we state the function or use of the device. For audiences such as customers, the use of the device is an attention-getting characteristic. Next we give our audience an idea of the overall appearance of the whole device and its dimensions. If the device is a new or unfamiliar one, we might compare it with a familiar object or identify the geometric shape: for example, "this device looks like a ballpoint pen and is cylindrical." In identifying the size of the device, we indicate height, width, depth, diameter, and so on. Finally, in listing the main parts of the device, we give our audience an idea of what is to come, in the order in which it will come.

For example, in the following introduction to a device description of a Zippo® cigarette lighter, the writer defines his subject, indicates its purpose, describes overall size and shape, and lists the main and subparts:

A cigarette lighter is a flame-producing device which uses both mechanical and chemical principles and is used almost exclusively for the lighting of cigars, cigarettes and pipes. The central characteristic of the lighter is that a simple hand motion can ignite and produce a continuous flame.

The Zippo® cigarette lighter is approximately the size of a small matchbox, though heavier. It is easily carried in the hand or pocket. The overall outside dimensions are 54 mm × 32 mm × 13 mm (height, width, and depth, respectively). The lighter will be described as if viewed sitting upright on a table and positioned so that it could be operated by a right-handed person. The lighter includes four principal parts:

1. Case: composed of the main case, cover, and hinge
2. Lighter body: composed of the fuel reservoir housing, wind shield, flint guide, and cover positioning mechanism
3. Fuel reservoir: composed of the fuel-retaining fabric and wick
4. Ignition mechanism: composed of the striking wheel, flint, and plunger assembly

Notice that the writer gives the audience a sense of orientation by stating that he will describe the lighter as if it were standing upright. In the introduction to a device description, we draw in words the overall frame of our visual picture.

**Body or Discussion.**   In the body or discussion of the description, we "take apart" the device to describe the parts. We verbally "detach" the parts to define them and describe their size, shape, function, material, location, and method of attachment to the whole and the other parts. In other words, we note many of the same kinds of characteristics about the parts as we did about the whole.

For example, in the device description of the lighter, the writer notes the definition, function, dimensions, material, and location and attachment of subparts of the case:

The case of the Zippo® cigarette lighter is a protective container for the inner components of the lighter. The two-piece case seals tightly enough to prevent evaporation of the lighter fuel and is made of stainless steel to protect the internal parts. The cover is also used to extinguish the lighter's flame by cutting off the oxygen supply.

The main case measures 32 mm × 38 mm × 13 mm, with a closed bottom and an open top. This thin shell serves to enclose the lighter body. The cover measures 22 mm × 32 mm × 13 mm, with a closed top and an open bottom. This shell protects the wick and part of the ignition mechanism. The open ends of the main case and cover meet each other when closed to seal the inside contents from the outside environment. Both the main case and the cover are fastened to a hinge at the left edge of their open ends. This hinge allows the cover to be swung up and back from the main case, exposing the upper portion of the lighter body.

If all the lighter parts had been composed of stainless steel, the writer could have stated this in his introduction; however, since materials vary, he must mention this characteristic when he describes each part. Since the shape of the lighter is the shape of the case, the writer did not repeat this characteristic in the description of the case.

**Describing Internal Parts.** In device description, we often describe what the eye cannot see, parts that are too small or parts that are enclosed inside other parts. We must choose carefully the words that convey to the audience a sense of location. Where exactly in relation to the parts we *can* see are the parts we *cannot*? How are the internal parts attached to the external parts or the whole device? We must give directions such as we would on a road map: what is to the right, to the left, up, down? Notice that when the writer of the lighter description moves "inside" his device, he stresses location and attachment to other parts and the part he has just described, the case:

The lighter body consists of the fuel reservoir housing, wind shield, flint guide, and cover positioning mechanism. This principal part serves to support and house all the integral parts of the lighter mechanism itself. The lighter body measures 50 mm × 36 mm × 12 mm, including the fuel reservoir housing and the wind shield. The lighter body is slightly smaller than the main case, allowing a nearly air-tight fit. It slips down into the main case far enough so that only the wind shield protrudes from the main case.

The fuel reservoir housing measures 34 mm × 36 mm × 12 mm and is made of stainless steel. The edges of this part are slightly rounded also. This thin metal shield encases the fuel reservoir.

Attached to the upper end of the fuel reservoir housing is the perforated wind shield, also stainless steel. The wind shield measures 16 mm × 16 mm × 13 mm. It resembles a small rectangular smokestack, or chimney, around the wick. The wind shield surrounds the wick and protects the flame from strong air currents which could blow it out. However, it is perforated to allow small amounts of oxygen to reach all parts of the exposed section of the wick. This arrangement facilitates efficient burning of the lighter fuel.

Two other parts are connected to the lighter body: the cover-positioning mechanism and the flint guide. The cover-positioning mechanism is situated to the left of the wind shield, near the hinge of the cover when the lighter body is inserted into the main case. This 13-mm-long steel piece pivots about a rivet which extends through two tabs on the lighter body. This spring-loaded mechanism restricts the cover to two positions: fully open or fully closed. The tension of the spring on the inside of the cover holds the cover in one of the two positions.

The flint guide is also fastened to the lighter body. This cylindrical, hollow, brass tube is 35 mm long with an outside diameter of 4 mm and an inside diameter of 1.5 mm. The flint guide is situated vertically inside the fuel reservoir housing. It runs parallel to the right face of the housing, 6 mm in from this face. It is also located midway between the front and back faces of the housing. The guide is soldered around the rim at its upper end to the top of the fuel reservoir housing, directly beneath the striking wheel. The flint guide remains open at both ends and directs the flint upward to come in contact with the striking wheel. . . .

In describing internal parts, we must give our audience a clear sense of orientation: where is the part in relationship to the other parts and the device as a whole? How is it attached to these other parts? We can easily note the material, shape, dimensions, and function of parts; the real challenge is in selecting words to describe their exact location.

*Conclusion.* In the conclusion of a device description, we must put the device "back together" for the audience by describing once more how the parts are attached to each other and to the whole. For example, after describing each part of a pencil in detail, we would say, "Thus a pencil is composed of lead which fits into a wood casing to which a metal band is attached which in turn holds a rubber eraser." We can put the device back together by reassembling the device verbally. For example, after describing the fuel reservoir and ignition mechanism, the writer reassembles the lighter and stresses again location, attachment, and function:

> The plunger assembly screws into the flint guide, holding the flint and forcing it against the striking wheel. The striking wheel and flint act together to create a spark when the lighter is operated. The fuel-retaining fabric and wick are contained inside the lighter body. They store and transport fuel up to the end of the wick, where it can be ignited by the spark. The lighter body surrounds the wick and the cotton fabric and prevents evaporation of fuel, as well as providing structure. The lighter body fits tightly inside the main case and the cover can be closed to protect the working component of the lighter.
>
> *David Dalrymple*

## Checklist for Device Description

The following is a list of those characteristics that usually appear in a device description:

I. Introduction
   A. Definition of the device
   B. Purpose or function of the device
   C. Overall shape and size of the device
   D. General description or list of the parts
II. Body: Description of each part of the device
   A. Definition of the part
   B. Purpose of the part
   C. Overall size and shape of the part
   D. Material makeup of the part
   E. Location of the part, relationship with other parts or the whole, and method of attachment of part
III. Summary description. Parts "back in place."

When dividing a device into parts, we should follow the natural separation of the parts. They should be distinguished by material, size, or shape. Our audience must gain knowledge of a complex object only from our words and perhaps a supporting graphic aid. Because in device description our audience follows our words like a map, we must balance our description of appearance and of attachment. We must choose a starting and a stopping place for our description and guide the audience from beginning to end. Most devices lend themselves naturally to our describing them from top to bottom, right to left, outside to inside, and such. Our "direction" must be consistent as we describe how one part is attached to the next. Like the differentia in definition, attachment is the most challenging part of description to write.

# Audiences and Purposes of Device Description

Since device descriptions are so specific and detailed, they are most appropriate for peer or lateral audiences, those audiences who want detailed accounts of research activities. We may write device descriptions for a concerned public or for potential users of a device as well as for supervisors who may approve a new device; however, most device descriptions are directed toward peers who want and can understand precise description. Let us look first at a few examples of device description directed outward or upward, before studying laterally directed device description.

## Outward-Directed Device Description

When we describe a device to the public, we select the most important details to include while following the overall pattern of device description. Our audience may be most interested in the function of our device and its parts, and least interested in the dimensions. The public wants to know how our developments will affect lifestyles. Therefore, rather than using precise measurements to describe every part, we might compare the part with a familiar object and stress the smallness or largeness of the part as well as its capabilities.

For example, in the following public relations release, the writers describe a new Bell system. They concentrate on the capabilities and functions of the parts. They include some precise dimensions but also compare the light source to a "grain of salt" and describe the photodetector as "tiny." Notice that the description moves from the general to the specific:

> In the Chicago system, a lightguide cable carries voice, data and video signals for about a half-mile between the Brunswick office building and an Illinois Bell central switching office (Franklin). Then, for another mile, between the Franklin office and a second central office (Wabash), the lightguide cable carries video signals and some of the voice signals normally carried between those two offices.

The video signals originate from a public PICTUREPHONE® Meeting Service room at Illinois Bell headquarters (across the street from the Franklin office), and from a customer installation in the Brunswick building. A conventional transmission link carries the video signals between the Franklin office and Illinois Bell's headquarters.

The lightguide cable contains 24 fiber lightguides manufactured by Western Electric. Only one-half inch in diameter, the cable is a fraction of the size of cables for lower-capacity systems now used to interconnect central offices in cities. A single pair of lightguides in the cable can carry 672 simultaneous conversations (at 44.7 megabit per second rate) or an equivalent mix of voice and various types of data signals. Although this capacity was judged sufficient for the volume of traffic carried, lightguide cables with 144 fibers and having a capacity of nearly 50,000 simultaneous conversations have been constructed and tested.

Each lightguide used in the system is connected at one end to a transmitter module containing either a solid-state laser or LED [light-emitting diode] light source, both smaller than grains of salt. (Though LEDs are less powerful than lasers, tests have shown they are adequate for relatively short transmission distances.) The LED and laser transmitter modules are interchangeable for the Chicago application.

The other end of each lightguide is connected to a receiver module containing a tiny photodetector, known as an avalanche photodiode, that converts light pulses to electrical signals. Terminal circuits then convert these signals into a format compatible with the nationwide telecommunications network.

The device description above stresses the purpose of the lightguide system and could be understood by a person with little technical background. When dimensions are included, their uniqueness is stressed; materials are used to emphasize ultimate capability, and technical terms that must be used are defined informally. Generally the system is described as being an improvement over old methods and an asset to the whole telephone system.

## Upward-Directed Device Description

Device description directed upward to supervisors and managers also stresses capability and usefulness. Detail of dimensions, shape, and attachment depends on the audience's interests and technical background. Often we describe a device for managers and supervisors to assure them that such a device would improve company operations. A supervisor or manager might approve our communication and then send it downward to other engineers and technicians.

For example, the following device description appeared in a technical report written to supervisors on how to "assist operators involved in tanker unloading operations in selecting the proper pumping combinations for each unloading situation." The major facilities of the dock and tanker unloading system are described:

The Seaway dock installation consists of one "T" head (Berth #1) dock for small tankers (from barges to 35,000 DWT tankers) and one finger pier with breasting dolphins designed for 120,000 DWT vessels. The outboard spot (Berth #3) of the finger pier has mooring dolphins designed for 85,000 DWT vessels and the inboard spot (Berth #3) has mooring dolphins designed for 50,000 DWT vessels.

Each unloading spot will consist of the following: Berth #1 will have one 16-inch marine unloading arm as the initial and ultimate configuration, a 30-inch unloading line from berth to meter station and four 12-inch turbine meters; Berth #2 will have two 16-inch marine unloading arms as the initial installation with three required for the ultimate configuration, a 36-inch unloading line from berth to meter station and two 16-inch plus one 8-inch turbine meters; Berth #3 will have one 16-inch marine unloading arm as the initial installation with two required for the ultimate configuration, a 36-inch unloading line from berth to meter station and two 16-inch plus one 8-inch turbine meters. This will give a metering capability at Berth #1 of 24,000 BPH (4 × 6000 BPH), Berth #2 — 27,300 BPH (12500 + 12500 + 2300 BPH) and Berth #3 — the same as #2. . . .

The control system provides pump unit protection for low suction pressure, high current demand, high temperatures in pump bearings, variable drive bearings, motor bearings and motor windings. The control system provides for modulation of the 1000 HP varidrive to control pressures and optimize the use of horsepower when this unit is in use. Manifolding is provided to enable selecting the pump combinations that are most efficient, with each pump combination able to take suction from any dock and discharge into either 30-inch line to Freeport terminal. Each 30-inch line operates completely independent of the other with any combination of pump units available to lines #1 and #2. Line #3 may use only pumping units #2 and #4.

In the device description above the writers stress the size and capabilities of the parts and how these capabilities contribute to the whole system. In upward-directed device description then we emphasize how a device may improve company operations; our description may be passed downward after the audience has approved the device. The writers include enough detail so that "parts" of the system can be ordered upon approval, and each part is justified by stating its exact purpose.

## Laterally Directed Device Description

Lateral audiences of device description want specific details about our activities to evaluate and even to reproduce them. Since at the least a lateral audience would contrast our device to ones already used, this audience needs exact dimensions and detail on material and attachment.

Often we describe new devices for a lateral audience. Although we must be exact especially in dimensions, material, and attachment, we can use symbols and abbreviations to convey this detail to a lateral audience. We write in mathematical code which a superior or a general audience might not be able or willing to interpret. For example, in the following device description of a new thermal seal that "insulates a gap that changes dimensions during use," the writers represent material, dimensions, and even capabilities by symbols and abbreviations:

A pair of fiberglass textile fibers, woven in a weave-and-tuft strip, are used to insulate a surface gap which is exposed to severe thermal stress. The strip is fabricated as a continuous, fibrous-pile composite which has low thermal conductivity, a working temperature range of from −454° to 2,000° F (−270° to 1,090° C), low load compressibility, and good inhibition of plasma flow.

The fibers are arranged in a brushlike orientation normal to the base. The sealant . . . consists of pile-woven yarns and filler yarns which form a mesh in one direction. Warp yarns, used as a backing, are looped around the filler yarns and pile yarns and form a mesh in a second direction. A binder prevents yarn separation in the backing and is secured to one of the fixed edges of the gap. The pile yarns contact the other gap edge (which is free to expand and contract while undergoing thermal stress). Fiber strip construction can be varied, depending on the type, density, and height of the fiberglass yarns used.

The yarns and stuffers are fabricated from either a composition fiber formed of 65 percent silicon dioxide, 25 percent aluminum dioxide, and 10 percent manganese dioxide, or a quartz fiber having a composition of 99.9 percent silicon dioxide. The composition is effective up to 1,400° F (760° C); the quartz, up to 2,300° F (1,260°C). After weaving, a binder of silicone rubber is applied to form an integral backing section. Pile density (excluding the backing) can vary from 9.5 lb/ft$^3$ (152 kg/m$^3$) to 27 lb/ft$^3$ (432 kg/m$^3$).

The audience of this device description could duplicate the device or at least decide to use it. The device described, unlike the cable system or the loading dock we discussed earlier, is just one part— a thermal seal—of a larger system or device. The audience is given a choice as to which materials can be used in the makeup of the device. The description then is directed toward experts, a lateral audience most qualified to understand and most interested in the precise detail of device description.

# Style in Device Description

## Numbers

In device description, we have to use numbers to specify dimensions and number of parts in a consistent and precise way. When we are specifying the number of parts or any other aspects of our subject represented by a noun, we use the figure for numbers of 10 or more and the written expression for the numbers one through nine. We would have

        10 bolts        23 parts        65 screws

and so on. But we would have

        nine legs        seven screws        one top

If in the sentence we have one or more numbers of 10 or more, to be consistent, we make all numbers figures:

        The top contains seven boards and nine metal strips.
        The top contains 7 boards and 12 metal strips.

We consider each sentence a unit and try to be consistent in using numbers in each sentence. In each sentence then, we look for the highest number and if that number is over nine, we use figures throughout the sentence.

## Units of Measurement, Time, and Quantity

Units of measurement, time, and quantity are always expressed in figures regardless of the size. Units of measurement are standard and allow us to measure other things. For example, we can measure a table in terms of inches, feet, millimeters, yards, and other units of measurement. Some units of measurement follow:

| | | |
|---|---|---|
| years | degrees | centimeters |
| months | seconds | horsepower |
| days | liters | megabit |
| P.M. | gallons | millimeters |
| A.M. | percent | watts |
| meters | acres | Angstroms |
| degrees | Hertz | kilovolts |
| inches | volts | microns |
| miles | amperes | nanoseconds |

Although figures are always used for units of measurement, time, and quantity, this rule does not affect the other rules for numbers:

The table has four legs and is 5 feet high.
The base is 5 inches long and 13 inches wide and rests on four
   pegs.

## Unit Modifiers

Unit modifiers are two or more words joined by a hyphen that serve as a "unit" to modify a noun. They function as an adjective to describe some characteristic of that noun. One of these words can be a number or a unit of measurement. The rules that we have discussed above apply to unit modifiers:

| | |
|---|---|
| 5-day week | 5-foot-wide entrance |
| 8-year-old boy | solid-state circuit |
| 12-foot pole | long-nosed pliers |
| two-man team | mirror-image halves |
| 15-man team | 2-millimeter diameter |
| ½-inch peg | 3-inch-diameter top |
| 7-second period | 19-percent raise |

One way to determine whether a group of words and numbers form a unit modifier is to ask if they tell us something about the noun that follows. For example, what kind of raise was it? a 19-percent raise. What kind of pliers? long-nosed pliers. What kind of period? a 7-second period. What kind of team? a two-man team. Note that "man" is not a unit of measurement. Men vary in size and shape and are not "standard"; therefore, we write out "two" because it is less than 10.

We must follow these rules on how to use numbers, units of measurement, time, and quantity, and unit modifiers throughout technical communication. The rules are especially important in device description, the pattern in which we must use precise dimensions. In addition to these rules, we usually do not begin a sentence with a figure, and if two numbers appear side-by-side, we use the figure for the unit of measurement, time, and quantity (e.g., four 2-foot beams and twelve 2-foot beams).

Again precise wording is most important in device description. Whenever possible, dimensions should be exact with the point of departure and terminal points given if not clear from context. The *point of departure* is where the measurement starts (e.g., *starting from the tip* of the eraser) and the *terminal point* is where the measurement ends (*to the end* of the unsharpened wooden casing, a typical pencil is 7½ inches long).

As we noted earlier, a sense of orientation and direction is essential in device description. We should make clear to our audience how they are "looking" at the device, from the bottom, top, side, from the inside out, and so on; and we should remind them of this direction or orientation throughout the description.

Using pronouns and synonyms may confuse our audience. We should refer to a part by the same noun, or noun and adjective, throughout. For example, if we identify a part as a "ceramic stem," we might confuse our audience if we called the part "the stem" later in the description.

Numbering parts helps an audience distinguish between the parts and anticipate which one we will describe next. However, once we begin to number the parts, we should continue to do so throughout.

Finally, we usually use the present tense of verbs in device description, and since we "freeze" our device in time and motion, we rely heavily on the verb "to be." Since device description demands such detail, our sentences should be short and concise. Using lists and subheads also breaks up long descriptive paragraphs.

# Graphic Aids in Device Description

We use graphic aids in device description to supplement our words rather than take their place. Only within the actual description can we convey to the audience with what familiar objects the parts can be compared, and the material, attachment, and function of the parts. The graphic aid might show the location of the parts and their dimensions, but the description itself contains the essential information.

The most useful graphic aids in device description are diagrams. A diagram of the entire device, a cutaway diagram to show the internal parts, and an exploded diagram to show parts detached are most commonly used in device description. Figures 4-1, 4-2, and 4-3 show the cigarette lighter described earlier in this chapter. Figure 4-1 shows the entire device and appears in the introduction

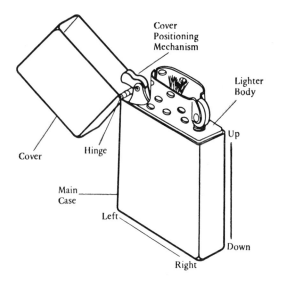

*Figure 4-1.*

to description; Figure 4-2 shows the internal parts of the device and appears in the body of the description; Figure 4-3 shows the internal parts of the ignition mechanism detached and appears in the section on that part. In diagrams then, as in the actual device description, we take apart a device to put it back together again for an audience.

In diagrams we should label the parts by the same name as in our description rather than by a letter or number code. These labels help the audience link our words with our illustrations. We should mention the graphic aid in our text the first time we describe any part represented.

Diagrams can also give our audience various views of our de-

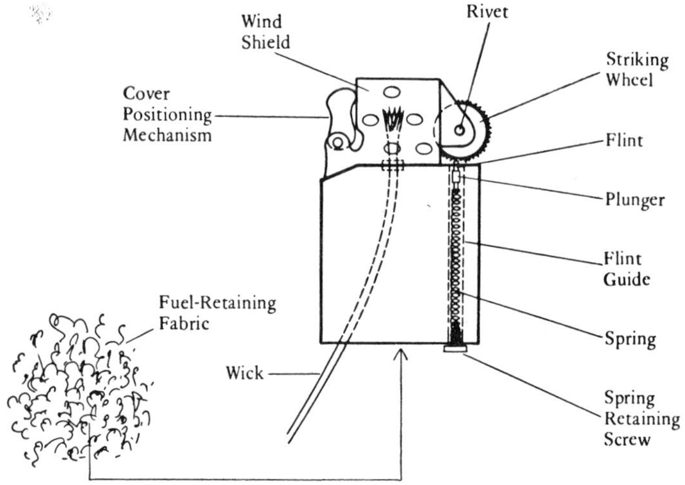

*Figure 4-2.*

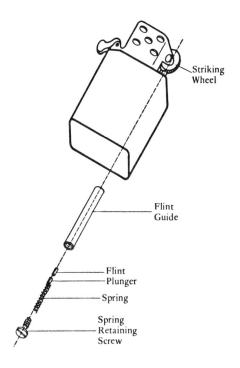

Striking
Wheel

Flint
Guide

Flint
Plunger

Spring

Spring
Retaining
Screw

*Figure 4-3.*

vice and support the orientation or direction of our description. We can give a front, side, top, or bottom view of the device and its parts. Diagrams then support device description by showing the location of parts and by illustrating hidden or nonvisible parts.

## Summary

Device description draws a verbal picture of a device for an audience by precise figures and words. Moving from the general to the specific, in device description we define a device, state its purpose, and describe its size and shape before defining each part and describing its function, material, size, shape, and location and method of attachment. Device description takes a device apart and then puts it back together for an audience.

While many device descriptions are directed laterally toward audiences most interested in and most qualified to understand precise dimensions and methods of attachment, sometimes we do describe a new device to the public or to a supervisor. The amount of detail and which characteristics we emphasize depend on our audience and purpose.

In device description, we must pay particular attention to figures and the written expression of numbers, units of measurement, time, and quantity, and unit modifiers. We must use precise wording and indicate orientation as we describe our device. Diagrams of the whole device, cutaway diagrams, and exploded diagrams support device description.

# Readings
## to Analyze and Discuss

# I
## The Surgical Replacement of the Human Knee Joint

One can best grasp what is involved in the replacement of a knee joint by considering the anatomy of a normal knee. . . . The joint includes two articulations, or movable parts, that transmit force and relative motion. One is the patellofemoral bearing surface between the patella (the kneecap) and the femur (the thighbone). The other is the tibiofemoral articulation between the tibia (the larger of the two long bones between the knee and the ankle) and the femur.

Three basic groups of muscles, under the control of the nervous system, provide the forces required for the functioning of the knee. The hamstring group and the gastrocnemius group contract to make the leg bend at the knee. The quadriceps group, utilizing the mechanical advantage provided by the patella, straightens the leg. These two movements are known respectively as flexion and extension.

The ligaments of the knee, composed mainly of parallel fibers of collagen, are pliant and flexible, so that they allow a considerable but nonetheless remarkably controlled freedom of movement. Along their length the ligaments form strong and relatively inextensible connections of bone to bone, thereby providing both stability to the joint and constraint of motion. The patellar ligament links the patella and the tibia and provides a sliding motion at the patellofemoral articulation during flexion and extension. Two pairs of ligaments, the collaterals and the cruciates, stabilize the tibiofemoral articulation.

The tibiofemoral articulation includes two crescent-shaped structures of fibrous cartilage: the medial meniscus (on the inside of the knee) and the lateral meniscus (on the outside). The upper surfaces of the menisci are concave, thus deepening the tibial plateaus for articulation with the femoral condyles, the knoblike structures at the bottom of the femur. Attachments to the relatively flat tibial surface allow motion of the menisci with respect to the tibia. (The common sports injury described as a "torn cartilage" involves one or both menisci.)

The medial and lateral femoral condyles are complex surfaces with continuously changing radii of curvature. These articular surfaces, in concert with the ligaments and the geometric form of the menisci and the tibia provide the stabilized and supple range of motion of the knee.

## Questions for Discussion

1. What are the audience and the purpose of the description above? What evidence supports your conclusion?

2. What typical characteristics of device description are included and which are excluded? Why did the writer make this choice?

3. What features of the human knee does the writer stress? Why? What types of diagrams might be included to support the description?

4. How might this description be rewritten to address a non-technical audience, such as a patient about to undergo surgery for knee-replacement?

# II
# USDA Quarantine Facility, Newark, Delaware

The 147-m² quarantine section consists of nine rooms, storage area, corridor, and three-section anteroom (Fig. 4-4). The building has a structural-steel framework on a reinforced concrete slab, and the roof is supported by steel decking. The two exterior walls are cement block, with aluminum and glass decorative panels. The internal walls, partitions, and ceilings are 1.3-cm-thick plasterboard attached by screws to galvanized-steel studs. The windows are sealed in anodized-aluminum frames by vinyl gaskets and cannot be opened. All rooms except 10, 11, and 12 have white walls, ceilings, and bench tops for easy visibility of stray insects. The four maximum-security isolation rooms (Q-I) have triple-glazed windows composed of double thermal glass for insulation, with the inside pane frosted for light diffusion, and a 0.65-cm acrylic outside pane to protect against breakage. These rooms also have gray, poured (seamless) epoxy floors and baseboards and specially finished walls, with all inside corners filleted to eliminate hiding places. The floors of rooms 5 through 9 are covered with gray or white 31-cm² vinyl tiles. The plastic-laminated bench tops in rooms 1 through 6 are matte white. The doors and frames are metal in rooms 1

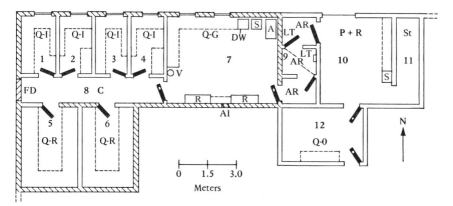

*Figure 4-4.* Floor plan of quarantine section. Q-I, Quarantine isolation. Q-R, Quarantine rearing. C, Corridor. Q-G, Quarantine general. AR, Anteroom. P & R, Packing and records. St, storage. Q-O, Quarantine office. DW, Dishwashing machine. A, Autoclave. R, Refrigerator. AI, Air intake. FD, Fire escape door. S, Sink. LT, Light trap. V, Central vacuum. High-security area (dark border), 97 m²; anteroom (buffer) area, 10 m²; nonsecurity area (quarantine office, packing and records, storage) 41 m².

through 9 and, except for the two swinging doors in room 9, all door perimeters are sealed with vinyl-magnetic gaskets that adhere tightly to the metal doors. All rooms except 10 through 12 have a white acrylic-latex semigloss enamel on the walls and ceilings to retard moisture and facilitate cleaning.

Tap- and cooling-water pipes, sewer pipes, and wires (electrical, telephone, and intercom) enter quarantine either from the concrete slab or the attic, which is walled off from all other sections of the building. All space around pipes and conduits penetrating through the ceiling from the attic is carefully sealed. Access to the attic for equipment maintenance or repair is through gasketed panels in the ceilings of rooms 8 and 11. All penetrations through the walls, except for the electrical wires and water pipes in room 12, are also sealed with silicone calking at both the inside and outside surfaces to prevent the escape of insects and eliminate hiding places. Rooms 7 and 10 have hot- and cold-water pipes for the sinks, autoclave, and dishwasher. Both sink drains have two successive traps to prevent insect escape. All rooms have three to eight 115-V electrical outlets divided among circuits that are arranged to prevent a single failure from affecting an entire room. A 230-V circuit serves the autoclave. . . . The main circuit breakers are located outside quarantine. Rooms 1 through 4 have vacuum connections to a central system that is powered by a unit in room 7. This system is used to clean these rooms and is suitably screened to prevent insect escape. Rooms 8, 10, and 12 have intercom telephones that can be used internally or for communication with other areas in the laboratory and the outside. Rooms 7, 8, and 10 have a separate intercom system that can be used to communicate with rooms 1 through 6 and other key locations in the laboratory. Persons in rooms 1 through 6 can reply to incoming messages without interrupting their work, but they must go outside into the corridor to initiate calls. Because the quarantine section is isolated from the rest of the building, a fire-warning gong is located in room 8. A fire-alarm pull switch is located next to the quarantine fire door. Emergency lights in rooms 7, 8, and 9 turn on automatically if an electrical power failure occurs. . . .

## Questions for Discussion

1. What are the audience and purpose of the description above? Support your answer with evidence.

2. What unusual characteristics does the writer include in the description that may not be included on our device description checklist? Why are these characteristics included?

3. What then are the basic differences between describing a device such as a cigarette lighter, an organ (the normal knee in Reading I), and a part of a building (Reading II)?

4. How would you describe the rooms in the quarantine facility if you wanted to describe the atmosphere of the building and to create sense (smell, touch, taste, sight, and sound) impressions for your audience?

5. Describe a room with which you are familiar—your dorm room, the school cafeteria, a classroom, or such—in a device description and then describe the same room to create the atmosphere of the room. What are the differences in audience, purpose, organization, and characteristics included in the two descriptions?

# *Exercises and Assignments*

1. **Problem:** Perhaps the most difficult object to describe is a common object. Because we assume that the audience is familiar with the parts, we are too selective in our description.

Assignment:

Read the following device description of a common object, the clothespin. Test the accuracy and completeness of the description by:

1. Examining the object as you read the description

2. Drawing the object from the description only

A clothespin is a small, hand-operated, clamping device usually made out of two wooden sticks held together in a V-like shape by a metal spring. It is normally used to hold clothing on a clothesline.

The two wooden sticks which make up the clothespin are identical in size and shape. Both are $3\frac{5}{16}$ in. long, $\frac{1}{4}$ in. wide (for $1\frac{11}{16}$ in. of their length), and $\frac{5}{16}$ in. across. For the final $1\frac{5}{8}$ in. of their lengths the sticks are tapered. The thick end of the clothespin is the clamping end, while the thin or tapered end is the handle end. On the inner surface of both sticks, near the clamping end, are two grooves which are used for gripping the clothes to the line. The first groove is about $\frac{1}{4}$ in. from the thick end, and the second, smaller groove is about $\frac{11}{16}$ in. from the same end.

A portion of the spring consists of three turns of wire wound into a helix. This portion rests between the two pieces of wood, fitting into semicircle shaped grooves in each stick (one in each stick) which are about $1\frac{11}{16}$ in. from the thick end of the clothespin. From each end of the helix, the wire extends at an angle of 45 degrees for a distance of $\frac{3}{8}$ in. until it reaches a U-shaped elbow. From these elbows come the lower part of the wire arms which run parallel to the upper arms for a distance of 1 in. They then bend to go horizontally across to the outside of the sticks, one arm resting in a groove which begins at a point $\frac{15}{16}$ in. from the thick end of the clothespin, and the other arm resting in a groove in the same position on the outside of the bottom stick. The arm resting in the groove in the top stick exerts pressure downward, and the arm resting in the groove in the bottom stick exerts pressure upward. Since the helix is between the two sticks, the pressure from the arms keeps the thick end closed and forces the thin end open, giving the clothespin its characteristic shape.

By squeezing the two sticks on the thin end together, you can force the thick end to open. This is the principle involved in using the clothespin to hang up clothes. When pressure is applied to the thin end, against the pressure exerted by the spring, the clamping end (thick end) will open. You then place the items you wish to clamp on the line, put the open thick end over them, and release the thin end. The thick end will then go back to its static, closed position, clamping the items firmly to the line.

*Paula Hugunine*

**2. Problem:**   Since device description creates a verbal picture of an object for an audience, the test of accuracy is whether that audience can draw the object from the description alone.

Assignment:

**1.** Write a description of a device that has at least three parts. Leave out the introductory definition and statement of purpose. Do not name your device within your description.
**2.** Exchange papers with another student. Attempt to draw the object the other student has described, while that student attempts to draw yours. Do not try to guess the object. Follow the description exactly and represent all errors in your drawing.
**3.** Exchange papers again. Discuss with your partner what you both left out of your drawings and how the descriptions might have misled the audience. Decide what worked well in the descriptions and what guidelines determine how to write successful device descriptions.

**3. Problem:**   One test of accuracy in device description is whether the audience can draw the object after reading the description. We can reverse this process and describe a device from a drawing alone. This exercise tests our ability to create a verbal picture of a device. In Figure 4-5 various views of a combination screwdriver are depicted. (All measurements are expressed in inches; Hex stands for hexagonal; the four "tips" are magnetic.)

Assignment:

Using Figure 4-5, write a complete device description of the combination screwdriver. Include a statement of audience and purpose.

**4. Problem:**   On-the-job assignments usually call for us to describe a device we have examined and support our description with graphic aids.

Assignment:

**1.** Write a device description of an object that has at least five parts.
**2.** State the audience and purpose of your description.
**3.** Support your description with graphic aids.

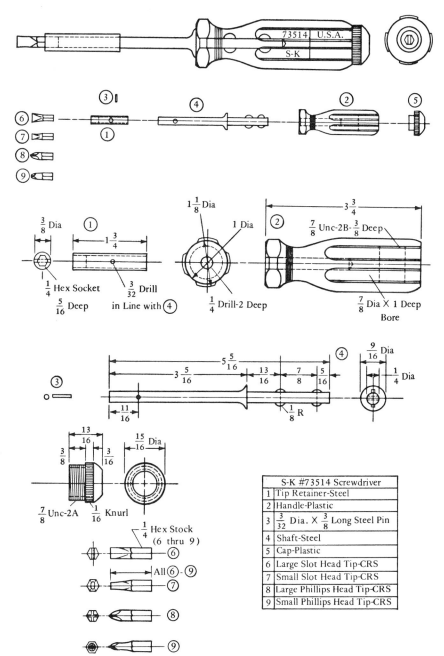

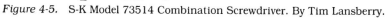

*Figure 4-5.* S-K Model 73514 Combination Screwdriver. By Tim Lansberry.

# PART THREE

## Presenting Multiple Subjects to an Audience

# 5

# Comparison-Contrast:
## Identifying Similar or Different Characteristics of Subjects

*Using simultaneous comparison and alternating, block, and implied comparison or contrast.*
*Choosing between or combining patterns of comparison-contrast.*
*Comparing by analogy.*
*Using comparison-contrast in introductions.*
*Comparing and contrasting alternative recommendations.*
*Reading and analyzing comparison and contrast written by others.*

After we have defined and described our subject for an audience, we can increase understanding of the subject by comparing it with or contrasting it to another. We can also introduce more than one subject by comparison-contrast. If our subject is a new one, we might compare it with a more familiar one. If we want to consider the best of several alternatives, we might contrast the qualities of one subject to those of another. We use comparison-contrast to present more than one subject or to emphasize the qualities of one subject in light of another.

Comparison identifies similar characteristics of two or more subjects that are usually thought to be different. Contrast identifies different characteristics of two or more subjects that are usually thought to be similar. The two techniques can be combined if the subjects have some similar and some different characteristics.

For example, in the following classical passage, Pavlov contrasts in detail the reflexes of the "bold" dog to those of the "cowardly" dog. While all dogs share similar characteristics, Pavlov identified differences upon which he based his study of the nervous system:

> Right from the very beginning of our experiments with dogs based on the method of conditioned reflexes we (like others) were struck by the different behaviour of the bold and the cowardly dogs. The former offered no resistance when led to experimentation; they remained quiet in the new experimental conditions, both when they were placed in the stands mounted on tables, and when certain apparatuses were attached to their skin and even placed in their mouths. When food was given to them by means of an automatic device, they began to eat it at once. Such was the behaviour of bold animals. But the cowardly animals had to be accustomed gradually to the procedure—a process which required days and even weeks. Another difference was observed when we began to elaborate conditioned reflexes in these dogs. In the first case the conditioned reflexes developed rapidly, after the application of two or three combinations; they reached considerable strength and remained constant, no matter how complicated the system of reflexes. In the second case, on the contrary, the conditioned reflexes were formed very slowly, after many repetitions; their strength increased at a very low rate, and they never acquired stability, being sometimes even at zero, no matter how considerably their system was simplified. It was,therefore, natural to assume that in the first dogs the excitatory process was strong, while in the second it was weak. . . .

Pavlov contrasted the same characteristics of the bold and cowardly dogs. We can compare and contrast subjects of equal importance or emphasize the characteristics of one subject in light of those of another, perhaps more familiar, subject.

# Patterns of Comparison and Contrast

## Simultaneous Comparison

When we compare our subjects simultaneously, we mention both subjects and a similar characteristic at the same time, usually in one sentence. We use this pattern when we want to mention many similar points, perhaps to convince an audience that two different subjects do indeed have a lot in common. Because we mention each subject simultaneously, we cannot go into depth on any of the common characteristics. We can only demonstrate the quantity of these similar characteristics. For example, the following paragraph is organized by simultaneous comparison. The subjects are sprinters and distance runners. Although they compete in different events, they have a lot in common. Notice that the author relies on the words "both" and "each" as well as the pronoun "they" to present his subjects simultaneously:

> Both sprinters and distance runners go through the process of getting "psyched" for a big race. Each spends weeks in preparation, training hard during the day, and getting plenty of rest at night. The night before the meet, both of them are sure to get to sleep very early, so they are able to awake in the early morning and start their mental, as well as physical, process of loosening up. During these morning hours, each of these individuals spends much time alone, concentrating on the skills and weaknesses of their opponents. Once they are at the track, both the sprinter and the distance runner know what they have to do. They both avoid their opponents for as long as possible, mentally building themselves up. Immediately before their respective races, both greet their opponents with a handshake of luck, hoping for a good, clean race from which they will emerge victorious.
>
> *Robert Slaski*

We use simultaneous comparison then to enumerate many similar characteristics of our subjects. Generally one characteristic is named in each sentence, and subjects can be represented by words such as "both" or "each."

## Uses of Simultaneous Comparison

Of course, to sustain simultaneous comparison for too long might overwhelm an audience. Simultaneous comparison can establish basic similarities which introduce an in-depth study of more important similarities or differences. For example, in the following passage, Loren Eiseley lists several characteristics shared by Henry David Thoreau and Charles Darwin. However, Eiseley's basic concern is with their essential differences: the practical, confident Darwin versus the subtle, elusive Thoreau:

> Both men were insatiable readers and composers of works not completely published in their individual lifetimes. Both achieved a passionate satisfaction out of their association with the wilderness. Each in his individual way has profoundly influenced the lives of the generations that followed him. . . .
>
> Both men forfeited the orthodox hopes that had sustained, through many centuries, the Christian world. Yet, at the last, the one transcends the other's vision, or amplifies it. Darwin remains, though sometimes hesitantly, the pragmatic scientist, content with what his eyes have seen. The other turns toward an unseen spring beyond the wintry industrialism of the nineteenth century, with its illusions of

secular progress. The two views, even the two lives, can be best epitomized in youthful expressions that have come down to us. The one, Darwin's, is sure, practical, and exuberant. The other reveals an exploring, but wary, nature.

Eiseley uses simultaneous comparison then to show his audience how much this philosopher and this scientist had in common before he explores the essential characteristics that separated them.

Although we can compare two subjects at the same time, we cannot contrast two subjects simultaneously. For example, the opposite of the word "both" is "neither"—still a term of comparison, not contrast.

## Alternating Comparison and Contrast

In alternating comparison and contrast we identify one characteristic of one subject, and then we compare or contrast immediately the same characteristic of the second subject. As with simultaneous comparison, we can mention many characteristics but not describe them in depth. Alternating comparison and contrast asks the audience to look from one subject to the other and back again several times. To make these transitions clear, we usually separate the subjects and characteristics compared or contrasted by sentence structure and punctuation.

For example, in the following paragraph, the writer compares the strategy of the long-distance runner with that of the sprinter. He uses alternating comparison, mentioning in one sentence a characteristic about long-distance running, in the next sentence a characteristic about sprinting, and so on, because he wants to impress his audience with the number of similarities:

> Although they are different types of races, the strategies behind running a long-distance race and a sprint are basically the same. In a long race, it is common practice to stay up near the front of the pack in the early portion of the competition. Similarly, the start is crucial to the sprinter, and often determines the outcome of the contest. Once a long-distance runner breaks through the initial stage of the race, he loosens up, and his stride begins to develop. This is also true with a sprinter; his short choppy steps convert into long, graceful strides soon after the first 40 yards of the race. Long-distance runners often finish the race with a strong kick, proving that the tiresome race has not worn them down. In the last 20 yards of a race a sprinter shows his true strength, searching for that last burst of power which will allow him to cross the finish line first. As you can see, the distance runner's strategy is very methodical and well-planned, just like that of the sprinter.
>
> *Robert Slaski*

In this example, the writer separates visually the characteristics of the sprinter from those of the runner by sentence structure. His audience sees the similarities between the two athletes in a balanced, side-by-side comparison.

## Uses of Alternating Comparison-Contrast

The preceding example demonstrates balanced alternating comparison and contrast. However, we can alter this pattern to sustain our audience's interest and to support our ideas. In the following example, Carl Sagan contrasts the effects of lesions in the brain's hemispheres. Although he uses the basic alternating pattern, he gives more information about the right hemisphere than the left, since the effects of the right are more difficult to grasp:

> The first evidence that these two modes of thinking are localized in the cerebral cortex has come from the study of brain lesions. Accidents or strokes in the temporal or parietal lobes of the left hemisphere of the neocortex characteristically result in impairment of the ability to read, write, speak and do arithmetic. Comparable lesions in the right hemisphere lead to impairment of three-dimensional vision, pattern recognition, musical ability and holistic reasoning. Facial recognition resides preferentially in the right hemisphere, and those who "never forget a face" are performing pattern recognition on the right side. Injuries to the right parietal lobe, in fact, sometimes result in the inability of a patient to recognize his own face in a mirror or photograph. Such observations strongly suggest that those functions we describe as "rational" live mainly in the left hemisphere, and those we consider "intuitive," mainly in the right.

Sagan modifies the alternating contrast pattern to emphasize new information about the right hemisphere in light of what is known about the left hemisphere.

We use alternating comparison and contrast when we want to look at several characteristics of each subject. Because we move from one subject to the next and back, our audience will not lose track of what characteristic we are comparing or contrasting. We can alter this pattern slightly to sustain interest or to emphasize one subject.

## Block Comparison and Contrast

When we want to describe in depth the most important characteristics of our subjects or when we cannot alternate readily these characteristics, we discuss one subject completely before going on to the second. This "block" pattern is also helpful when we discuss interrelated characteristics, ones that do not fall naturally into categories. However, when using block comparison and contrast we have to link the two subjects frequently or our audience will forget what we said about the first subject while reading about the second.

For example, in the following paragraph, the writer contrasts a distance runner's workout to a sprinter's. Since the characteristics he mentions are interrelated steps in the training or workout process, he uses block contrast. Notice that he refers to the runner in the second half of his paragraph so his audience will not forget either subject of the contrast:

> The training process for a long-distance race is quite different from a sprinter's workout. Distance runners train for consistency, trying

to run ten quarter-miles, in close sequence, at a constant pace. Their prepractice warm-up is often lengthy and just a sign of things to come. They train over long periods of time; a single practice can last up to four hours. On the other hand, sprinters' workouts are short, concise, and much more intense. Their warm-up is brief, consisting mainly of stretching their muscles. They devote much of their time to working on starts, trying to master the use of starting blocks. Unlike that of the runner, the main portion of their workout is devoted to speed-work, running a mere 110 yards at a time. Although the distance runner trains over long periods of time, it is the sprinter who must be able to handle the sting of a concentrated practice.

*Robert Slaski*

Because the writer discusses the subjects in block pattern, the audience can contrast the complete workout process of each.

## Uses of Block Comparison-Contrast

Sometimes, an audience must choose between the subjects compared or contrasted, whether to buy or use one subject rather than the other. In block comparison-contrast we can best present objectively each choice and let the audience compare and contrast specific detail according to need. For example, in the following passage from an engineering report, the writer contrasts open-pit to underground mining. Which process an audience will choose depends on individual mining conditions. The writer uses the block pattern since the audience consists of professionals who need few guidelines to follow the contrast, and since the steps of the process are not separable but interrelated:

**Open-Pit Mining.** While each open-pit mine is unique in its detail, the basic approach is the same for all such mines.

The orebody is first exposed by stripping off the overburden. Depending on the hardness of the rock, this is accomplished by large scrapers, bulldozers or combinations of shovels, front-end loaders and haulage trucks. . . .

One of the key considerations in open-pit mining is grade control. Uranium deposits often consist of interlayered bands of ore and waste material, sometimes as little as one foot thick. In order to maintain a relatively constant grade of material being fed to the mill, the operator must constantly decide whether material being mined should go to the dump (waste), the leaching pads (low-grade ore) or the blending stockpiles (ore grade). This requires near-continuous sampling of the active mining areas and frequently involves blending or mixing of higher grade ore from one area of the mine with lower grade ore from another area. In many mines, each truckload of material is scanned with scintillometers to determine its uranium content and the driver directed to specified dumping areas based on the readings.

**Underground Mining.** Here again, each mine varies in specifics, but a typical underground mine is characterized by a series of shafts through which ore, men and materials, and fresh air are moved. These shafts, which can range from a few feet up to 25 feet in diameter, are sunk either with large-diameter shaft-boring equipment or by drilling, blasting and hoisting the rock out. . . .

Underground mining has a number of obvious economic and environmental advantages, but they are often offset by the need for extensive mine ventilation systems. This is particularly true of uranium operations because of the presence of radioactive decay emissions. Strict regulations concerning radon gas levels and individual cumulative exposures have been established, and are monitored closely by State and Federal agencies.

Although when using block comparison or contrast we refer to both subjects frequently to help an audience keep them in mind, sometimes we let our audience draw detailed comparisons or contrasts themselves, as did the writer of the mining contrast. Our comparison or contrast may seem more objective when presented in the block pattern.

## Implied Comparison and Contrast

When we devote most of a comparison or contrast to one of our subjects, usually the less familiar one for our audience, we can imply rather than state the comparison or contrast. Our audience may be so familiar with one subject that to describe it would be unnecessary, even boring or insulting. In implied comparison or contrast, we mention both subjects at the beginning and end of our discussion, but devote the body of the discussion to the less familiar subject.

For example, Isaac Asimov "implies" a comparison-contrast between primordial and contemporary earth. Since his description of the conditions of primordial earth that may have caused spontaneous generation is new and speculative, he emphasizes earth *then* rather than earth *now*. He assumes that his audience is familiar with the characteristics of earth now:

One obvious difference between modern Earth and primordial Earth, for instance, is that modern Earth has life and primordial Earth had not. Any chemical substance that arose spontaneously on Earth today and that was approaching the level of complexity where it might be considered as protolife would surely be food for some animal and would be gobbled up. In the primordial and lifeless Earth, such a substance would tend to survive (at least, it would not be eaten) and would have a chance to grow still more complex and to become alive.

Then, too, the primordial Earth might have had an atmosphere that was different from the present one.

This was first suggested in the 1920s by the English biologist John Burdon Sanderson Haldane (1892–1964). It occurred to him that coal was of plant origin, and that plant life obtained its carbon from the carbon dioxide of the air. Therefore, before life came into being, all the carbon in coal must have been in air in the form of carbon dioxide. Furthermore, the oxygen in air is produced by the same plant-mediated reactions that absorb the carbon dioxide and place the carbon atoms within the compounds of plant use.

It follows, then, that the primordial atmosphere of the Earth was not nitrogen and oxygen, but nitrogen and carbon dioxide. (This sounds even more logical now than it did when Haldane suggested it, since we now know that the atmospheres of Venus and Mars are made up largely of carbon dioxide.)

Furthermore, Haldane reasoned, if there were no oxygen in the air, there would be no ozone (a highly energetic form of oxygen) in the upper atmosphere. It is this ozone that chiefly blocks the ultraviolet light of the Sun. In the primordial Earth, therefore, energetic ultraviolet radiation from the Sun would be available in much larger quantities than it is now.

Under primordial conditions, then, the energy of ultraviolet light would serve to combine molecules of nitrogen, carbon dioxide, and water into more and more complex compounds that would, finally, develop the attributes of life. Ordinary evolution would then take over, and here we all are.

What could be done on the primordial Earth, with lots of ultraviolet, lots of carbon dioxide, no oxygen to break down the complicated compounds, and no living things to eat them, could not be done on present-day Earth with its dearth of ultraviolet light and carbon dioxide and its overabundance of oxygen and life. . . .

Asimov then proves that the characteristics that make primordial earth different from contemporary earth support the theory of spontaneous generation. Since his audience lives on contemporary earth, he concentrates on primordial earth. He "implies" the contrast but lets his audience complete the contrast patterns in its mind. In implied comparison or contrast we let the audience balance the two subjects in its mind while we concentrate our description on the new or less familiar subject.

## Analogy

Analogies compare two or more very different subjects. We use them to get an audience's attention, to make something new and perhaps abstract more concrete, or to persuade an audience to look differently at a familiar subject. Analogies are comparison, not contrasts, because they identify similar characteristics of subjects that are usually considered different.

For example, in the following analogy, the writer helps her audience understand the complicated forces in a "black hole" in space:

> The ability of a black hole to withhold a photon of light is like the ability of the earth's gravity to stop a baseball from escaping its atmosphere.
>
> *Barbara Strollo*

An analogy is like a metaphor or simile in literature. It can announce the writer's approach to or attitude toward a subject. In analogy we draw together two subjects of different classes to make one more vivid or concrete for our audience.

For example, in the following analogy, the writer explains how tall plants survive in their environment by drawing an analogy between a tree and a high-rise building:

Structural support is a major requirement for any dwelling. Without support the smallest vegetable enterprise protruding vertically an inch or more into the atmosphere would be toppled by the strong force of gravity and the shearing force of winds. In the absence of foraging animals or of competition for space and light with other plants, a prostrate herb might (and some do) survive admirably; however, the 300 to 400 million years of land plant evolution have resulted in a tremendous species diversity that makes survival difficult for all plants because of intense competition for raw materials and real estate. Although diminutive plants still have a place in nature, the space race for plants, like that for humans, clearly resulted in construction of the vegetable skyscraper. Such high-rise architecture contains increased office space (leaves) and also towers above the herbaceous congestion that competes for the available sunlight. . . .

The specialized supportive cells in plants have cell walls constructed much like reinforced concrete. Both contain long solid rods that are carefully oriented within a surrounding matrix of congealed material. In a plant cell wall the solid rods are long thin needles of cellulose called microfibrils; in concrete the rods are steel. For added strength in a given direction, the cellulose microfibrils are aligned in parallel array within a single layer of the wall. Most thick walls have several layers, and the different layers often have the cellulose microfibrils oriented at different angles for multidirectional strength. (The same engineering principle is often used when pouring concrete.) The cellulose rods are embedded within a gluelike matrix consisting of amorphous carbohydrate (hemicelluloses and pectins), protein, and often lignin (a complex polymer especially characteristic of wood). But whereas concrete is a congealed solid (of cement and sand or gravel), the matrix of the cell wall is a colloid with a thick gellike consistency, held together as a mesh by chemical bonds between the various constituents. The wall's fluid nature imparts a desirable resiliency to the structure—thus wood is less brittle than most concrete. The quality of the final product is attested to by the thousands of uses to which we put lumber and wood products. . . .

By sustaining an analogy throughout our discussion, we can help our audience understand a new and complicated subject. We can also provoke interest by comparing our subject with one usually not considered the same.

## Comparing and Contrasting Many Subjects

Discussing similar or different characteristics of *more* than two subjects is a common communication task. For example, in many upward-directed reports we compare and contrast several alternatives before recommending one. In these discussions, we have to not only maintain a balance between the subjects and their characteristics, but also guide the audience by frequent reference to all our subjects. When we discuss many subjects, the block and alternating patterns of comparison-contrast are useful in maintaining this balance and guiding the audience. We might take each subject in turn in the block pattern, while using the alternating pattern to introduce each block structure.

# Audiences and Purposes of Comparison-Contrast

Because we use comparison-contrast often in technical communication to introduce a subject and to recommend an alternative, let us look more closely at these specific situations and their audiences.

## Comparison and Contrast in Introductions

To place a communication within a context for an audience, we often compare or contrast what has happened before our study or what has happened in another location. By comparing and contrasting subjects in time or space, we can set the stage for our discussion. We can show why our communication is important, new, or progressive.

We can also use comparison and contrast to introduce a communication in which we report progress on a project. Because our audience may want to know what we have accomplished in a certain time period, we can contrast what we have done most recently to what we did in the immediate past; we might contrast work completed this month to work completed last month. We compare and contrast the same characteristics of each subject, in this case, in each time period.

## Comparison and Contrast in Recommendations

Often we compare and contrast alternative actions before recommending one to an audience. For example, we might compare and contrast various types of copying machines according to cost, time, structure, difficulty in operation, durability, and manufacturer's reputation, before we recommend one type to our audience. Comparing and contrasting alternatives is especially useful in communications directed upward.

If we want our audience to arrive at our conclusions only after considering all options, we can compare and contrast all aspects before recommending one option. However, if we sense that our audience is unable or unwilling to keep all aspects in mind until the end of our discussion, we can include our recommendations within the comparison-contrast pattern. Usually we compare and contrast alternatives in upward-directed communication for an audience who will approve our recommendations or make the final decision. We usually use the block or alternating pattern or a combination of the two.

For example, in the following comparison and contrast of cost and structural requirements for lighting methods in a building to be renovated, the writers use the block pattern to discuss each option. Since the writers are engineers addressing a management audience interested in the specific requirements of each option, the writers include their recommendations in their discussion of each method:

The following methods of light control have been considered:

a. Local Room Control–Low-Voltage Switching. This method has previously been used in the building in limited areas and because of the constraints of the stainless steel grid is the only feasible method of local switching with this ceiling system. The estimated cost for one low-voltage switch installation including wiring for one average size room is $250.

Based on allowing minimum of one LV switch/room, the total estimated cost of using LV switching throughout the building is $500,000 (2,000 rooms with individual switches).

Based on first costs, and allowing for some energy savings, this method is not economical and is not recommended for use throughout the building. However, it can and should be used for special areas where local switching is required.

b. Local Room Control–High-Voltage Switching (265V). This method is practical only if a totally new ceiling system was to be installed (all stainless steel grids removed) and the selected type of partitions permitted installation of wiring and standard switch box. Under those conditions, the cost of one local switch would be approximately $50.

c. Central Control–Manual or Automatic. The original design of the 265/460-volt lighting panels serving the general office fluorescent lighting allowed space for future addition of remote control switches. The addition of these switches and required control wiring would permit central on and off operation of lights from one location either manually or automatically (if a computer based automation system is added). We estimate the initial cost of a remote control system to be $275,000 and the resultant annual energy and labor saving conservatively to be approximately $65,000. On that basis, we recommend the installation of the remote control system for general lighting.

Since the audience of this communication will choose the method best suited to budget and renovation plans, the writers include the basis for each recommendation but leave the final decision up to the audience.

Often we use comparison-contrast to discuss alternative recommendations in a communication. If we want to lead our audience to a logical final conclusion, we might compare and contrast all aspects of the options before recommending one. If our audience is unwilling or unable to keep in mind many aspects of our comparison-contrast, we might include our recommendations within the comparison-contrast. Using comparison-contrast in introductions creates a context that helps our audience understand and appreciate our discussion, while using comparison-contrast in discussing recommendations shows our audience how we arrived at our conclusions.

# Style in Comparison and Contrast

As we discussed each comparison, we determined certain stylistic requirements for each. In simultaneous comparison-contrast, words such as "both" and "each" link the two subjects. In alternating comparison-contrast, transitions guide the audience from one subject to the next. In the implied and block pattern, frequent reference to both subjects reminds the audience of the characteristics compared and contrasted.

## Linking Subjects

Let us look more closely at how to link subjects in comparison-contrast by transitions and repetition. In the following alternating contrast, the writer notes the many differences between peanut butter and jelly. The repeated and transitional words appear in italics:

> One of the staples of the American diet has been the *peanut butter and jelly* sandwich. The connection between *peanut butter* and *jelly* is spontaneous worldwide. It is really remarkable that these two items so closely connected have nothing in common.
>
> *Peanut butter* is a thick, carmel-colored substance, *while jelly* comes in a variety of colors and has a lesser viscosity than *peanut butter*. The difference in the textures is a very interesting characteristic of *peanut butter* and *jelly*. *Jelly* does not stick to the roof of your mouth, and rarely do you rip your bread when attempting to spread *jelly* on it which you invariably do with *peanut butter*. *Jelly* can come in a variety of *flavors while peanut butter* has the same *flavor* jar after jar. It is *also* a known fact that *peanut butter jars* open more easily than *jelly jars*. *Peanut butter* seems to be the more widely recognized of the *two*. This is probably because the advertising for *peanut butter* can use elephants and "peanut men" *while jelly* has no such gimmicks and faces stiff competition from jams, preserves, and marmalades. Probably *another* reason for the popularity of *peanut butter* is the brand name. *Peanut butter* comes under catchy labels such as Skippy or Peter Pan *while jelly* comes under less spectacular labels such as Smucker's or Welch's. *But* despite these differences, *peanut butter* and *jelly* do go well together, and they will continue to be found between pieces of bread throughout America and the world.
>
> *David Henning*

Even in this light, brief contrast, the writer repeats his subjects and their characteristics often so that the audience can keep them in mind. Transitions such as "while" and "also" guide the reader back and forth between subjects in the alternating pattern.

## Ordering Subjects

When reading comparison-contrast in the block pattern, our audience must remember the characteristics we mention about the first subject to see the similarities and differences when we discuss the second subject. Although transitions signal when we change subjects, we can help our audience remember all the compared and contrasted characteristics by mentioning these characteristics in the same general order. In the following example, the characteristics are numbered to show how the same writer balanced and ordered the characteristics of his subjects, peanut butter and jelly, in the block contrast:

> So often in this world we take things for granted that it is rare that we look beyond popular beliefs to see the truth behind the misconceptions. The case of peanut butter and jelly is an illustration. The public is so "brainwashed" into thinking that peanut butter and jelly go together that they never realize that peanut butter and jelly have nothing in common.
>
> Peanut butter has only one flavor (1) that varies little from jar to jar. It has a thick texture (2) and is carmel colored (3). Because of this thick texture, one has to be very careful when spreading peanut butter on bread (4) or large holes in the bread will result (5). The texture also makes peanut butter stick to the roof of your mouth (6). Peanut butter comes in jars that are easy to open (7). Perhaps this is part of the reason that peanut butter is popular throughout America (8). The use of elephants and "peanut men" in advertising may be another reason (9). Another successful advertising tool has been such brand names as Skippy and Peter Pan (10) as opposed to duller brand names of other products on the market.
>
> Jelly, on the other hand, comes in many colors (3) and flavors (1). It is easier to spread (2,4). You rarely hear of anyone ripping bread while trying to spread jelly on it (5), and you never hear of jelly sticking to the roof of anyone's mouth (6). As a rule, jelly comes in jars that are harder than usual to open (7). Jelly does not have cute advertising gimmicks to boost sales (9), since elephants don't eat jelly. This reason, and the fact that jellies use brand names such as Smucker's and Welch's (10), make jelly the unsung partner in peanut butter and jelly sandwiches.
>
> *David Henning*

Because the writer discussed the same characteristics in generally the same order, the audience can follow easily his contrast.

In these models, we see the importance of transitions, repetition, order, and balance in comparison-contrast.

*which is more effective?*

# Graphic Aids in Comparison and Contrast

When we compare and contrast statistics or data, we should support our discussion with graphic aids. The two types of graphic aids that best compare and contrast data are tables and bar charts.

Tables, which we can use to compare and contrast any number of variables, enable an audience to see similarities and differences in subjects quickly and easily. For example, in Table 5-1, a writer who compares and contrasts numerically controlled and conventional machines presents four variables. In the table he summarizes the similarities and differences discussed in his communication.

Bar charts compare and contrast a constant and one or more variables. While we use tables to summarize data for an audience, we use bar charts to represent visually that data. For example, in Figure 5-1 the flank wear of coated and of uncoated tools is contrasted. Four different sets of carbide tools with nose radii of 1/64,

**TABLE 5-1  Conventional and Numerically Controlled Machines**

|  | Conventional | N/C |
|---|---|---|
| Time to make 1 part | 52.8 min. | 8.4 min. |
| Cost per bolt | $23.00 | $ 14.71 |
| Hourly machine rate* | $26.00 | $105.00 |
| Number produced | 30 pieces | 400 pieces |

*The hourly machine rate was determined by dividing the time to make 1 part into 60 min. and multiplying it by the cost per bolt.

*Martin Knox*

$\frac{1}{32}$, $\frac{3}{64}$, and $\frac{1}{16}$ inches are contrasted to four sets of titanium carbide coated tools with the same range of nose radii. The bar represents visually the test findings for the audience.

   Tables and bar charts in particular can present graphically the similarities and differences in the subjects we are comparing and contrasting.

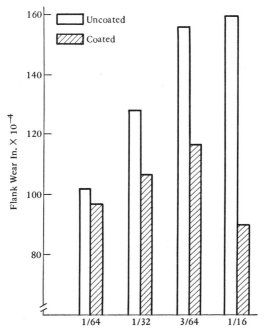

*Figure 5-1.*  Tool wear after 40 minutes. By Brian K. Lambert. Copyright © 1976 Society of Manufacturing Engineers. By permission of publisher.

# Summary

We have seen that comparison describes similar characteristics of two or more subjects usually thought to be different, while contrast describes different characteristics of two or more subjects usually considered similar. Four basic patterns of comparison-contrast, simultaneous, alternating, block, and implied, help us organize our thoughts. We use the simultaneous and alternating patterns to discuss briefly many characteristics, the block pattern to compare and contrast in depth interrelated characteristics, and the implied pattern to emphasize one subject when our audience is very familiar with the other. These patterns can be combined and varied according to our audience and purpose. Analogy compares two or more subjects that are very different to create a vivid picture for the audience or capture attention.

In upward- and downward-directed communication, we often use comparison and contrast to introduce subjects and to discuss alternative recommendations. When we compare and contrast, we link our subjects by transitional and repeated words while presenting the similar and different characteristics in a balanced order. We can also support comparison and contrast graphically by tables and bar charts.

# Reading to Analyze and Discuss

## I

Assuming that the brain and the computer are both machines, how are the two to be compared? The exercise is interesting. Computers are invented by man and are therefore thoroughly understood, if human beings can be said to understand anything; what they do not know is what future computers will be like. The brain was created by evolution and is in many important ways not understood. Both machines process information and both work with signals that are roughly speaking electrical. Both have, in the largest versions, many elements. Here, however, there is an interesting difference. For cells to be manufactured biologically appears to be reasonably simple, and neurons are in fact produced in prodigious numbers. It seems to be not so easy to increase the elements of a computer, even though the numbers are expanding rapidly. . . .

A still more important difference is a qualitative one. The brain is not dependent on anything like a linear sequential program; this is at least so for all the parts about which something is known. It is more like the circuit of a radio or a television set, or perhaps hundreds or thousands of such circuits in series and in parallel, richly cross-linked. The brain seems to rely on a strategy of relatively hard-wired circuit complexity with elements working at low speeds, measured in thousandths of a second; the computer depends on programs, has far fewer elements and works at rates at which millionths of a second are important. Among brain circuits there must be many devoted to keeping evolution going by means of competition and sex drives. So far the computer seems free of all that; it evolves by different means.

## Questions for Discussion

1. What subject does the writer stress in the comparison-contrast above? The brain or the computer? Why?

2. Name the characteristics that the brain and the computer share. Do not share. Can you name characteristics the writer does not mention?

3. What patterns of comparison and contrast does the writer use to discuss his subject?

4. What are the audience and purpose of the passage?

# Exercises and Assignments

**1. Problem:** You have been asked by your employer to conduct a workshop for new employees on technical communications. You want to explain the similarities and differences between the audiences of technical communications, in particular the audiences of upward- and downward-directed communications. To make sure that your presentation is balanced and well organized, you need to outline the characteristics of both audiences.

Assignment:

1. Analyze your own audience and purpose of the workshop.
2. List the characteristics of the audiences of upward- and downward-directed technical communications that you would compare and contrast in your workshop presentation. Make sure that you mention the same type and number of characteristics for both audiences.

**2. Problem:** As we saw in Chapter 1, we can compare and contrast the style and attitude toward the subject (tone) of communications on the same subject.

Assignment:

1. Review any two of the articles on the fall of the Soviet satellite Cosmos 954 at the end of Chapter 1. Notice again the stylistic characteristics and tone of each article. Decide whether the two articles are basically different or similar.
2. Write a communication that compares or contrasts the two articles for an audience familiar with the event but not with the specific articles. Use either the alternating or block pattern of comparison or contrast. You may add details about your audience and purpose if you wish.

**3. Problem:** You are writing an article in which you introduce a technical or abstract subject to a high school audience. To capture your audience's attention and to make your subject more concrete, you want to open your article with an analogy.

Assignment:

1. Choose a subject with which you are familiar. Choose either a technical subject or an abstract subject, such as beauty, loyalty, or love.
2. Write the opening paragraph to your article in which you compare your subject with one that is usually considered different but that would be familiar to your high school audience. Develop your paragraph to illustrate fully your analogy.

**4. Problem:** As with device description, a good test of the accuracy of a comparison or contrast is whether our audience can

*Figure 5-2.*   Photograph by Faye Serio.

visualize our subject from our words alone. We can also reverse this process by representing an illustration solely by our words.

## Assignment:

**1.** Study carefully Figures 5-2 and 5-3.

**2.** Write a communication in which you compare some aspects of the two illustrations. You may choose any audience and purpose, and you may use whichever of the three patterns of comparison is most appropriate for your purpose.

**3.** Write a communication in which you contrast some aspects of the two illustrations. You may choose any audience and purpose, and you may use whichever of the four patterns of contrast is most appropriate for your purpose.

**5. Problem:**  As we have studied, each one of the four patterns of comparison-contrast is appropriate for a certain audience and purpose. Simultaneous comparison and alternating comparison-contrast allow us to identify many characteristics of two or more subjects, block to discuss in depth a few interrelated characteristics, and implied to emphasize one subject.

*Figure 5-3.* Photograph by Faye Serio.

## Assignment:

**1.** Choose two subjects in the same general class with which you are familiar. Decide whether the subjects are basically different or similar.

**2.** Write one paragraph either comparing or contrasting the two subjects for each of the patterns of comparison-contrast that follow:

(a) simultaneous comparison or alternating comparison or contrast

(b) block comparison or contrast

(c) implied comparison or contrast

Change your audience and purpose for each of the three paragraphs you write.

**6. Problem:** As we studied in this chapter, we use comparison-contrast often to introduce subjects or to discuss alternative recommendations. Comparison-contrast in introductions creates a context for understanding the discussion to come while comparison-contrast of alternatives leads an audience to our conclusions and recommendations.

## Assignment:

1. Choose a technical or semitechnical subject with which you are familiar.
2. Place yourself in a hypothetical situation and analyze your audience and purpose. For example, you might be an engineer assigned to investigate the purchase of a new piece of equipment. You would be writing an upward-directed communication to report on the results of your project. Or you might be a supervisor in a branch office of a bank. You would be writing an upward-directed communication to the general manager which compares and contrasts the benefits of one 30-minute coffee break per day for tellers and the benefits of two 15-minute coffee breaks per day for tellers.
3. Write the introduction to your communication. Use the most appropriate one of the four patterns of comparison-contrast.
4. Write the conclusion to your communication. Use the most appropriate one of the four patterns of comparison-contrast.

# 6

# Causal Analysis:
## Understanding the Causes and Effects of Subjects

*Using single-cause—single-effect, multiple-causes—single-effect, single-cause—multiple-effects, and sequential causal analysis.*
*Designating a cause by analyzing effects.*
*Predicting effects of occurrences.*
*Using causal analysis to recommend how to control or change causes and effects.*
*Proving an event or theory cannot account for a cause (reverse causal analysis).*
*Reading and analyzing causal analyses written by others.*

One way that we understand and control our environment is by causal analysis. In causal analysis, we look backward to find the causes of an occurrence, or we look ahead to predict or control the effects of an upcoming event. Looking backward increases our understanding of our environment while looking ahead helps us control it.

We convey the steps and results of our causal analysis to an audience. We may want our audience to accept our ideas and recommendations because of their consequences or causes. For example, by causal analysis, we might convince an audience that a new computer would increase profits and save time, or that the old computer is causing decreased profits and lost time. We might observe a new inflationary trend and attribute it to recent governmental spending, or we might dismiss past criticism of an antipollution policy as the cause of decreased land development and identify lack of skilled labor as the real cause.

If an occurrence always causes the same effects, we regard these effects as the result of a natural or scientific "law." For example, when the temperature of water is reduced to 32° F., it freezes. If two events are so closely linked that *whenever* one occurs so does the other, this cause-effect relationship becomes a natural law. However, usually circumstances, such as time and location, make causal analysis speculative. We find *sufficient* cause for observed effects, or *predict* what will occur should an event take place.

Within classical scientific literature, we find these same two rhetorical strategies: finding sufficient cause for effects and predicting consequences of occurrences. For example, in 1809, Lamarck explored the theory of evolution by looking back to the causes of certain changes in species. By analyzing evolutionary changes in many species, Lamarck speculated that new life-styles and environments caused species to develop new shapes and faculties:

> We have seen that the disuse of any organ modifies, reduces and finally extinguishes it. I shall now prove that the constant use of any organ, accompanied by efforts to get the most out of it, strengthens and enlarges that organ, or creates new ones to carry on functions that have become necessary.
>
> The bird which is drawn to the water by its need of finding there the prey on which it lives, separates the digits of its feet in trying to strike the water and move about on the surface. The skin which unites these digits at their base acquires the habit of being stretched by these continually repeated separations of the digits, thus in course of time there are formed large webs which unite the digits of ducks, geese, etc., as we actually find them. In the same way efforts to swim, that is to push against the water so as to move about in it, have stretched the membranes between the digits of frogs, sea-tortoises, the otter, beaver, etc.

Thus Lamarck sought the causes of observable changes in species.

On the other hand, in 1798, Thomas Malthus speculated that because population increased in a "geometric ratio" and food supply in an "arithmetical ratio," we would eventually starve:

The population of the island [Britain] is computed to be about seven millions; and we will suppose the present produce equal to the support of such a number. In the first twenty-five years the population would be fourteen millions; and, the food being also doubled, the means of subsistence would be equal to this increase. In the next twenty-five years the population would be twenty-eight millions, and the means of subsistence only equal to the support of twenty-one millions. In the next period the population would be fifty-six millions, and the means of subsistence just sufficient for half that number. And at the conclusion of the first century the population would be one hundred and twelve millions, and the means of subsistence only equal to the support of thirty-five millions; which would leave a population of seventy-seven millions totally unprovided for. . . .

While Lamarck looked backward into time to find the causes for changes in the species, Malthus looked into the future and predicted the consequences of overpopulation. Thus in causal analysis, we look for sufficient cause of the effects we observe as did Lamarck, or predict the future consequences of events as did Malthus.

# Single-Cause—Single-Effect Analysis

If we attribute one cause to one effect or predict a single occurrence will have only one consequence, we must be sure that we are not oversimplifying matters. When we identify only one cause of an effect or when we predict one effect of an occurrence, we must make clear under what conditions we make our statement or prediction.

When we identify a single cause and effect, we may really be analyzing the main or the most important cause and effect. We may argue that other effects or causes are not as, or are no longer, relevant. Usually we give evidence that our cause or effect is the most important. For example, in the following passage, the author identifies a single cause for declining fertility in the underdeveloped world. While other studies have argued that socioeconomic factors influence the number of children people have, this author finds statistical evidence that the main cause is simply the desire for fewer children:

> The major reason for the declining fertility, however, is apparently simply that people want to have smaller families. In 10 of the 15 countries more than half of the married women of reproductive age reported that they want no more children; in six of those countries the figure was more than 60 percent. One analysis has shown that the desire to have no more children is substantial in all socioeconomic groups; for women with four children the difference between urban and rural women, between educated and uneducated ones and between those married to men with high-level occupations and those married to men with low-level ones is less than 10 percentage points in most of the countries. This suggests, according to Kendall, that the desire to stop having children "may be less influenced by socioeconomic development" than has been thought. . . .

The writer identifies the main or most relevant cause of the effect—
fewer children. He does so by eliminating all other possible causes
or stressing that they are not as important and by supporting the
main cause by statistics.

When we identify a single cause of a single effect or predict one
consequence of an event, we must tell our audience under what
conditions we made our analysis or give evidence that we have found
the major cause or effect.

## Multiple-Cause—Single-Effect Analysis

Often we find that many factors cause a single or major effect. These
causes may occur at about the same time, and their interaction
produce the basic effect. For example, in the following passage the
author found many reasons why industries move from central cities.
Land is not available, or it is too expensive. Because industrial operations themselves depend on one-story plants, firms must expand
outward rather than upward. All these conditions, occurring about
the same time and in the same location, cause industries to leave
cities:

> The limited availability and high cost of central city land as compared
> with suburban and nonmetropolitan land do not significantly inhibit
> stores and offices from locating in central cities because they can and
> do economize on land through multistory development. As continuous
> process and automatic material handling have made single-story
> plants the norm, however, new industry has been discouraged from
> locating in central cities by land-related factors. Many cities cannot
> provide land in lots large enough for modern industrial plants, and
> even when such land is available, it costs several times as much as
> comparable land in outlying areas. Similarly, many central cities can-
> not provide additional land for plants, already located in central cities,
> that wish to expand their operations. In either case, central cities lose
> industry to jurisdictions with more abundant or less expensive in-
> dustrial land.

All these causes—lack of space, high land costs, the need for single-
story buildings—occurring at the same time contribute to industry's
moving from the cities.

## Single-Cause—Multiple-Effect Analysis

One occurrence or event may cause many effects. These effects may
be related or unrelated, be predictions or surprises, be of varying
importance and degree, and occur at the same or different times.
However, usually the effects are related in some way. For example,
in the following passage, one company analyzes all the effects of the
1975 Federal Energy Policy and Conservation Act:

> As a result of the establishment of these minimum fuel economy
> standards, major efforts are being made by automobile manufacturers

to improve fuel economy by reductions in size and weight, by use of smaller engines, advanced ignition systems, and innovations to increase engine efficiency, and by reducing transmission losses, wind drag, and rolling resistance. The use of diesel engines in passenger cars and light trucks is being expanded because of the inherent efficiency of the diesel and because the volumetric heat content of diesel fuel is about 10% higher than gasoline. The economy standards are also promoting interest in advanced alternate power plants for vehicles. . . .

All these effects occurred about the same time in related areas: alternate power, diesel fuel, smaller engines, and so on. Although the Energy and Conservation Act did not "cause" diesel fuel, the act did stimulate further research in this area. Because one occurrence can have many consequences, we try to account for all of them.

## Sequential Causal Analysis

Often one occurrence will produce an effect that in turn causes something else to happen. This chain can continue indefinitely with each new occurrence causing another. Because predicting the final effect in the sequence is difficult, the end result may be opposite of the one we desire or envision. For example, in the following passage, the writers analyze the final effect of higher minimum wages. While the government increases this wage to support certain workers, the last link in the chain of effects may mean unemployment for those same workers:

> While government is extremely important in determining whether science and technology are used to make the Nation more productive, much of what the government has been doing that affects productivity has been the result of indirect, unintended, and often accidental effects. For instance, the government decrees a higher minimum wage. One consequence is that suddenly it becomes economically sensible for many buildings to go to automatic elevators. In some old buildings elevator operators are fired, become unemployed . . . and less productive! The automatic elevator industry expands and becomes more productive. Looking at our GNP a little later we note that a higher fraction of it is now in technological products—all of those added automatic elevators. Whatever happens in the buildings with automatic elevators takes place now with less people employed to make it happen, that is, less these same dismissed elevator operators—so the productivity of labor in that building is increased. We have substituted capital expenditures for labor expenditures. The science and technology advance represented by the design of the automatic elevator was not what was planned when the minimum wage law was passed. The act was really intended to provide a higher income to the elevator operators, who, it turns out, are now unemployed instead and have a lower income. . . .

The writer feels that the government failed to predict the final effect of a higher minimum wage. The consequences along the chain, such as building owners finding it more economical to convert to automatic elevators, lead to the reverse effect that the government wanted. The workers become unemployed.

In analyzing multiple causes and effects, often occurring about the same time, we must account for all the causes and all the effects. In sequential causal analysis, we must try to identify the last effect in an often long chain to see if that final effect is the one we predicted and wanted.

All four basic patterns of causal analysis—single cause and effect, multiple causes, multiple effects, and sequential cause-effect—require us to analyze completely all possible causes and effects. Causal analysis is like tracing all the clues in a mystery. Before we can find the real criminal, we must identify and account for all the important clues. In causal analysis, we must be sure to identify the real cause and the final effect. Let us now examine when and how to use causal analysis in technical communication.

# Causal Analysis in Technical Communication

## Finding Causes

Analysis is the process we use when we separate a subject into its parts to see how they interact to form the whole subject. As we have seen, causal analysis helps us see how causes have contributed to an event. When we conduct an experiment, we are usually looking for the cause of the effects we observe in the laboratory or environment, or we want to see how a new discovery will affect an old process. For example, we might search for the cause of a disease by examining the symptoms or effects in patients, or we administer a new drug to the patients to see how it will affect these symptoms.

When we search for the cause of observed effects, we usually examine and eliminate a number of possible causes. Once we find the probable cause, we then see whether under controlled conditions that cause will indeed produce the same effects. When we report the results of our experiment to an audience, we explain our procedure step-by-step. We represent our causal analysis by recreating for the audience all our actions and thoughts.

For example, in the following passage, the writers explain how they isolated the cause of Legionnaires' disease. Although the writers suspected that they had found the cause of the disease, they had to see whether the disease had caused certain antibodies to form in the patients' serum:

> *Establishing Cause.* We did that by testing the serums of the Legionnaires' disease patients for antibodies to the newly isolated organism. When a patient's serum is found to contain antibodies specific for the antigen molecules on a given organism, it is said to be "seropositive"; one can assume then that the patient has been exposed to that particular organism (or possibly one with similar antigens) at some time. If the patient's antibody level is found to have risen in the course of convalescence from a given illness, "seroconversion" has been dem-

onstrated and it is very likely that the illness was caused by the particular organism.

When we tested serums from the Legionnaires' disease patients for antibodies to the newly isolated organism by indirect fluorescent-antibody tests . . . , more than 90 percent of the serum specimens taken during convalescence turned out to be seropositive. In more than 50 percent of the cases—most of those for which suitably timed serum specimens were available—we were able to demonstrate seroconversion, indicating that the patients had recently been infected by this particular bacillus.

The writers isolated the cause after studying the effects of symptoms, and then confirmed the cause by creating certain effects under laboratory conditions. When reporting the results of their experiment, they explained their analysis step-by-step.

## Predicting Effects

On the other hand, when we predict for an audience the consequences of an occurrence or new condition in the environment, we must describe concretely how things stand now and how we think they will change. We describe our predictions as concretely as possible to make our audience care about these consequences. For example, in the following passage, the writers predict the effect of "uncertainty," caused by unpredictable government attitudes, interventions by critics, and capital scarcity, on the nuclear industry. To convince the audience that the writers' predictions are probable, they first describe present conditions and then each step of the sequence of effects: delay in new construction, demand for government assurances, and finally the nationalization of the nuclear industry and generating facilities. Here the writers analyze the most extreme consequences to alert their audiences to look ahead:

> Forty-year equipment lifetimes, slow staff turnover, and continued similarity of product have made the utility industry conservative. Since utilities get their rewards not from building any one type of plant, but rather from providing electricity to their customers, they delay new construction in the hope that dwindling energy supplies will create a public consensus on what types of facilities ought to be built.
>
> Nuclear power, the most controversial of the energy-generating technologies, is given up first. Indeed, the principal nuclear vendors may opt out of the nuclear business even earlier than the utilities. . . . None of the vendors has more than 25 percent of its business in nuclear power, and unless uncertainty is reduced enough to spur new plant orders, the majority will likely drop out. Recovery would prove difficult. The vendors, having built up a considerable industrial infrastructure yet in vain, would surely demand assurances of government financial and policy support before reentering the market. The continuing controversial nature of nuclear power would likely make such assurances impossible.
>
> If the current trend continues, and we see no changes in prospect, only the federal government may remain willing and able to build new nuclear plants. The result would be a *de facto* nationalization of the

nuclear generating business. But this is not all. As more becomes publicly known concerning the hazards inherent in other energy sources—coal, for instance—they, too, are likely to become highly controversial. The subsequent proliferation of uncertainty might well cause the utilities to opt out of constructing *any* new generating facilities and concentrate instead on the less risky business of distributing electricity to their customers. By default, America would be left with a nationalized electricity-generating business. . . .

The writers describe as concretely as possible present and future conditions.

When we analyze effects to identify a cause, we eliminate all other causes and then, if possible, see if our cause will indeed produce these effects under the same conditions. On the other hand, when predicting effects of a new occurrence or condition, we must describe completely and concretely how present conditions will be changed. We recreate for the audience our analysis, and we draw a vivid and sometimes extreme picture of future effects. Only by helping the members of our audience visualize these consequences will we convince them that our prediction is correct and that they should accept or change the cause of these consequences.

## Recommending How to Control or Change Causes and Effects

Many of the examples of causal analysis we have just discussed persuade the audience in some way. By concrete description or by recreating experiments, we convince the audience that we have identified the major causes or effects of occurrences. Often in technical communication, our causal analysis leads to recommendations. We identify the cause of a problem and recommend a way to eliminate it. We analyze the effects of a process and recommend ways to control, increase, or decrease these effects.

For example, in the following passage, the writers identify three causes of dust problems in highway construction: excavation, hauling, and compaction. Because the other two causes are more closely regulated, hauling appears to cause the most dust pollution. Having identified this cause, the writers recommend ways of reducing dust pollution:

> Embankment construction primarily involves three steps: excavation, hauling, and compaction. The excavated earth normally contains sufficient moisture, so that little, if any, dust results. The compaction process, under most state specifications, requires a certain optimum moisture content to achieve maximum compaction, so that this operation, too, is not a dust generator. The hauling process can generate dust, because the top surface is subject to drying and, when dried, offers a dust-producing source to passing vehicles. Winter months and days during and after rainfall constitute a significant portion of the year, such that the opportunities for drying are only a fraction of the construction season in many regions. Where dust production is likely, traffic can be banned, speeds can be reduced, or the hauling road or detour road can be watered. During extended periods of

drying, water appears to be inefficient because the effects may only last a few hours. Consequently, use of hydrophilic materials, such as calcium chloride, is more sensible. Further, on heavily traveled haul roads, cutback asphalts and emulsions may be used for dust mitigation. On the basis of the recent trend in the cost of asphalt products, emulsions have become the more favored choice. Where dust results from the spillage of embankments onto paved roads, frequent cleaning through scraping, sweeping, and housing is the best method for eliminating the potential for dust. . . .

The writers recommend hydrophilic materials, asphalts and emulsions, and frequent cleaning as the best methods to control dust caused by hauling. We can support any recommendations or suggestions we make to our audience by analyzing the causes of a problem and then analyzing the effects of our recommended solutions.

## Reverse Causal Analysis

Although usually we convince an audience by our thoroughness that we have identified the cause or the effect of a situation, sometimes we can prove the "real" cause or effect by eliminating all others. This process of elimination is a kind of "reverse" causal analysis. We can use this reverse causal analysis to disprove old or incorrect theories before we present our own. When we ask an audience to approach an old subject in a new way, we might prove that previously identified causes do not account sufficiently for all effects.

For example, Marvin Harris, the cultural anthropologist, dismisses all alternative theories of the origin of war, war as solidarity, war as play, war as human nature, and war as politics, before proving his own theory. In the following passage, Harris uses reverse causal analysis to convince his audience that war is *not* human nature:

War as human nature. A perennially favorite way for anthropologists to avoid the problem of specifying the conditions under which war will be regarded as a valuable or an abhorrent activity is to endow human nature with an urge to kill. War occurs because human beings, especially males, have a "killer instinct." We kill because such behavior has been proved successful from the standpoint of natural selection in the struggle for existence. But war as human nature runs into difficulties as soon as one observes that killing is not universally admired and that intensity and frequency of warfare are highly variable. . . . The Pueblo Indians in the Southwest of the United States, for example, are known to contemporary observers as peaceful, religious, unaggressive, cooperative peoples. Yet not so long ago they were known to the Spanish governor of New Spain as the Indians who tried to kill every white settler they could get their hands on and who burned every church in New Mexico, together with as many priests as they could lock inside and tie to the altars. . . .

Harris's theory of war is that "war and female infanticide are part of the price our stone age ancestors had to pay for regulating their populations in order to prevent a lowering of living standards to the bare subsistence level." War is not part of human nature but a reaction to culture and environment. Before Harris could convince his audience that his theory was sound, he had to show that other theories were inaccurate by reverse causal analysis.

We can use causal analysis to eliminate all other explanations before we discuss our own. In reverse causal analysis, we eliminate all *insufficient* causes or effects.

## Graphic Aids in Causal Analysis

Graphic aids are useful in causal analysis when we are analyzing highly related causes and effects that occur within a limited time period. We can best show the effects of multiple causes or the causes of multiple effects in tables if these causes or effects are related.

For example, in Table 6-1, the effects of voltage variation on

**TABLE 6-1.   Voltage Variation Effects on Computer Operation**

| Duration (seconds) | Percent of Voltage Variation | | | |
|---|---|---|---|---|
| | 0-10% | 10-20% | 20-60% | 60% & over |
| $\frac{1}{1000}$ to $\frac{1}{100}$ | no effect | errors | errors | errors |
| $\frac{1}{100}$ to $\frac{1}{10}$ | no effect | errors | errors & shutdown | errors & shutdown |
| $\frac{1}{10}$ to $\frac{1}{2}$ | no effect | errors | errors & shutdown | errors & shutdown |

computer operation are illustrated. The "varying conditions" are the causes of errors in operation, that is, the duration and percentage of voltage variation. The table helps the audience see the results of causal analysis. Again to use such a table, we would be analyzing related causes.

In Table 6-2, the effects or problems in using a certain tool, the spade drill, are illustrated along with the probable causes of such effects. Here we have a variety of related effects assigned to several related causes. Again the table summarizes for the audience the likely causes of related effects.

Graphic aids then, especially tables, are most useful in causal analysis of related, multiple causes and effects.

**TABLE 6-2. Troubleshooting the Spade Drill**

| | Problems | Most Likely Cause |
|---|---|---|
| Chips: | clockspring shape | Feed too light |
| | powdery | Feed too light |
| | broken into small pieces | Speed too high, feed too light |
| | long, stringy | Speed too high, feed too light |
| | deep purple/blue color | Speed too high, inadequate coolant |
| Hole: | off location, crooked | Starting with too-long holder, surface too rough or sloping, blade sharpened incorrectly |
| | exit burr | Blade dull, inadequate coolant at breakthrough, insufficient machine rigidity |
| Blade: | edges chipped | Feed too light, edges burned during sharpening, hesitant starting, holder too long (drill walking), use of lead hole |
| | edges burned | Speed too high, feed too light, using flood coolant, using dull blade. . . . |

# Summary

When we look back to identify the cause of effects we have studied and when we look ahead to predict the effects of an occurrence, we use causal analysis. Causal analysis helps us understand and control our environment. Generally we can use one of four patterns or a combination of them to investigate causal analysis: single-cause—single-effect; multiple-cause—single-effect; single-cause—multiple-effect; and sequential cause and effect. In all cases we must be sure to identify sufficient cause and complete effect.

Often we use causal analysis to support recommendations in technical communication. We might recommend how to change or control effects or how to eliminate the cause of a problem. We can use reverse causal analysis to disprove other theories. Graphic aids such as tables support an analysis of related causes and effects. We will look at analysis again when we study argument and persuasion.

# Readings
## to Analyze and Discuss

## I

Richard Smith
Smith Products, Inc.
New City, NY 14610          Subject: Parker #1
                           West River Road
Dear Richard:
    I visited the subject location yesterday to check their Dayton alarm installation.
    They have a new LD25 alarm mounted directly outside the manager's office. The alarm is connected to a 578F switch on the back entry lever and a similar switch in series connected to the garbage room door. They reported that they were getting sporadic alarm from the unit. Although no one would go through the door, the alarm would go off. Also, at times, someone would go through the door, but the alarm would not sound.
    I checked both switches and found that they were operating as designed. Next I observed the entire operation but could find no real problem. Consequently, I inserted a new GF board in the LD25 alarm. The GF board contains all of the logic and switching for this unit. If there is an intermittent short in any of the components, it would be located here. I then gave the manager full instructions on how to use the alarm and explained all time delays.
    Since the alarm did not malfunction while I was on the job, I can only assume that an intermittent action may have been caused somewhere within the GF board that I replaced.
    If you have any further complaints, please feel free to contact me.

Sincerely,

John W. Snyder

## Questions for Discussion

    1. What are the audience and purpose of the communication above? What is the subject analyzed?
    2. Identify all the causes and effects discussed in the communication. What pattern or patterns of causal analysis do you find?
    3. Does the causal analysis help the writer and audience change or control the environment? How? Does the writer eliminate all other possible causes or effects?
    4. What will the writer do if the alarm fails again in the future? What means of analysis might he use?

# II
# On Magic in Medicine

Medicine has always been under pressure to provide public explanations for the diseases with which it deals, and the formulation of comprehensive, unifying theories has been the most ancient and willing preoccupation of the profession. In the earliest days, hostile spirits needing exorcism were the principal pathogens, and the shaman's duty was simply the development of improved techniques for incantation. Later on, especially in the Western world, the idea that the distribution of body fluids among various organs determined the course of all illnesses took hold, and we were in for centuries of bleeding, cupping, sweating, and purging in efforts to intervene. Early in this century the theory of autointoxication evolved, and a large part of therapy was directed at emptying the large intestine and keeping it empty. Then the global concept of focal infection became popular, accompanied by the linked notion of allergy to the presumed microbial pathogens, and no one knows the resulting toll of extracted teeth, tonsils, gallbladders, and appendixes: the idea of psychosomatic influences on disease emerged in the 1930s and, for a while, seemed to sweep the field.

Gradually, one by one, some of our worst diseases have been edited out of such systems by having their causes indisputably identified and dealt with. Tuberculosis was the paradigm. This was the most chronic and inexorably progressive of common human maladies, capable of affecting virtually every organ in the body and obviously influenced by crowding, nutrition, housing, and poverty; theories involving the climate in general, and night air and insufficient sunlight in particular, gave rise to the spa as a therapeutic institution. It was not until the development of today's effective chemotherapy that it became clear to everyone that the disease had a single, dominant, central cause. If you got rid of the tubercle bacillus you were rid of the disease.

But that was some time ago, and today the idea that complicated diseases can have single causes is again out of fashion. The microbial infections that can be neatly coped with by antibiotics are regarded as lucky anomalies. The new theory is that most of today's human illnesses, the infections aside, are multifactorial in nature, caused by two great arrays of causative mechanisms: (1) the influence of things in the environment and (2) one's personal life-style. For medicine to become effective in dealing with such diseases, it has become common belief that the environment will have to be changed, and personal ways of living will also have to be transformed, and radically. . . .

There is a recurring advertisement, placed by Blue Cross on the op-ed page of *The New York Times*, which urges you to take advantage of science by changing your life habits, with the suggestion that if you do so, by adopting seven easy-to-follow items of life-style, you can achieve eleven added years beyond what you'll get if you don't. Since today's average figure is around seventy-two for all parties in both sexes, this might mean going on until at least the age of eighty-three. You can do this formidable thing, it is claimed, by simply eating breakfast, exercising regularly, maintaining normal weight, not smoking cigarettes, not drinking excessively, sleeping eight hours each night, and not eating between meals.

The science which produced this illumination was a careful study by California epidemiologists, based on a questionnaire given to about seven thousand people. Five years after the questionnaire, a body count was made by sorting through the county death certificates, and the 371 people who had died were matched up with their answers to the questions. To be sure, there were more deaths among the heavy smokers and drinkers, as you might expect from the known incidence of lung cancer in smokers and cirrhosis and auto accidents among drinkers. But there was also a higher mortality among those who said they didn't eat breakfast, and even higher in those who took no exercise, no exercise at all, not even going off in the family car for weekend picnics. Being up to 20 percent overweight was not so bad, surprisingly, but being *underweight* was clearly associated with a higher death rate. . . .

You have to read the report carefully to discover that there is another, more banal way of explaining the findings. Leave aside the higher deaths in heavy smokers and drinkers, for there is no puzzle in either case; these are dangerous things to do. But it is hard to imagine any good reason for dying within five years from not eating a good breakfast, or any sort of breakfast.

The other explanation turns cause and effect around. Among the people in that group of seven thousand who answered that they don't eat breakfast, don't go off on picnics, are underweight, and can't sleep properly, there were surely some who were already ill when the questionnaire arrived. They didn't eat breakfast because they couldn't stand the sight of food. They had lost their appetites, were losing weight, didn't feel up to moving around much, and had trouble sleeping. They didn't play tennis or go off on family picnics because they didn't *feel* good. Some of these people probably had an undetected cancer, perhaps of the pancreas; others may have had hypertension or early kidney failure or some other organic disease which the questionnaire had no way of picking up. . . .

# *Questions for Discussion*

1.  What are the past causes for diseases that we believed in the past? Why was tuberculosis a "paradigm" or example of one theory of the causes of disease?

2.  What do some people believe now causes disease? Why have we developed this new mode of thinking?

3.  What pattern of causal analysis was the California study based on? Did this study analyze all possible causes and effects?

4.  How else may the results of the survey be analyzed? Does the alternative analysis better explain the results? How does the writer "turn cause and effect around"?

5.  What are the audience and purpose of this passage? The tone or attitude toward the subject?

6.  What other areas of science or technology are often oversimplified or exposed to faulty causal analysis?

# Exercises and Assignments

**1. Problem:** Study Figure 6-1. Speculate on or simply invent the causes of the situation depicted. Use your imagination as much as possible.

**Assignment:**

Write a brief communication in which you analyze the cause of the situation depicted in the illustration. Be as thorough and as concrete as possible. Choose an appropriate audience and purpose.

**2. Problem:** As assistant manager of a local department store, you are in charge of 70 sales people who work on 2 shifts. On Mondays, Tuesdays, Wednesdays, and Saturdays, all employees work from 9:30 A.M. to 5:30 P.M., but on Thursdays and Fridays, employees work either of 2 shifts: 9:30 A.M. to 5:30 P.M. or 1:00 P.M. to 9:00 P.M. On these 2 days, the early shift takes a lunch break at 1:00 P.M. when the late shift arrives. The first half of the late shift takes a dinner break at 5:00 P.M., the second half at 5:30 P.M. You have staggered the shifts so that half the shift is on the floor at any one time. Lately you have noticed that the early shift has been leaving for lunch up to 15

*Figure 6-1.* Photograph by Faye Serio.

**117**

minutes early, which means that some departments are without a salesperson from 12:45 P.M. to 1:00 P.M. Also in the evening, the first half of the late shift has been coming back from dinner break up to 20 minutes late. Although you provide a cafeteria in the basement of the store, most of your employees eat in a restaurant nearby or bring their meals to work. While the second half of the late shift may be annoyed at having dinner 20 mintues late, the longer lunch periods are more disruptive to business. You must analyze the cause(s) of the two situations, determine whether they are related, and find a solution.

**Assignment:**

1. List all the possible causes of the situations described. You may add detail about both if you wish.
2. Choose the most likely cause(s) on your list. Write a communication to your immediate supervisor in which you analyze the causes of the situations, determine whether they are related, and suggest a solution.
3. Write a communication to all employees in which you describe briefly the cause of the situations and give instructions that will eliminate the problem.

**3. Problem:** You are a consulting engineer for your local town planning board. Because the town is on a direct truck route to neighboring rural areas, and because of a new shopping area downtown, traffic in town has increased 100 percent in the last 10 years. The main street through town is heavily traveled and is lined with the major stores and businesses. At the next town board meeting, you are going to suggest that the town widen the main street from the present 2 lanes to 4 lanes, eliminate parking along the busiest 4 blocks of the main street, and build a free public parking area 2 blocks from the main street. You know that representatives from the teamsters' union, the shop and business owners, and the town beautification committee will attend the meeting. To prepare for any objections, you must analyze all the consequences of the changes.

**Assignment:**

1. List all possible consequences of the changes. Be sure to consider all audiences. You may add detail about the situation if you wish.
2. Write separate communications explaining the consequences and recommending the change to go to each of the following audiences before the meeting:
   (a) the chairperson of the local town board
   (b) the chairperson of the town beautification committee
   (c) the head of the local teamsters' union
   (d) the local shop and business owners

**4. Problem:** Choose a theory or situation with which you are familiar or are willing to do some research. It can be a local or school situation, such as long lines at registration; a natural phenomenon, such as a plant or animal disease; a physiological condition, such as tooth decay or headaches; a philosophical or moral problem, such as the high divorce rate; or a technical problem, such as pollution. Choose a situation that needs to be controlled or changed in some way.

Assignment:

Addressing an appropriate audience, write a communication in which you (1) analyze the causes of the situation or problem you have chosen and recommend change or control, and (2) support your recommendation by predicting the consequences of adopting or not adopting your recommendation. You might want to support your communication with graphic aids and eliminate all other recommendations by reverse causal analysis.

# 7

# Classification and Partition:

## Grouping Subjects According to a Basis

*Classifying and partitioning on the basis of structure, function, material, or cause.*

*Choosing an appropriate basis for various audiences and purposes.*

*Dividing and subdividing completely in classification and partition.*

*Using a chart to check or clarify classification and partition.*

*Using classification to introduce a communication.*

*Using classification to explain on-the-job procedures.*

*Using classification to recommend alternatives.*

*Reading and analyzing classification and partition written by others.*

When we present many subjects to an audience, we often group our subjects according to common characteristics. This grouping helps us to organize our communication, and our audience to make sense of our discussion. The common characteristic we choose is the *basis* for our grouping. We might arrive at this basis by comparing or contrasting characteristics of our subjects or by identifying their main cause or effect. Or, we might derive our basis from the differentiae that define our subjects or from the main parts identified in device description.

Whether we arrive at our basis by definition, description, comparison-contrast, causal analysis, or some other means, we use the basis to group many subjects or, in effect, make sense out of what could be chaos. For example, if we describe various types of trees, we would confuse an audience if we started with pines, maples, and oaks and then went on to firs and willows. All trees either retain or lose their leaves seasonally; they are evergreen or deciduous. This characteristic is then a basis for grouping and discussing trees.

Scientists have used classification to study the plant and animal kingdom since Linnaeus published his book, *Systema naturae,* in 1735. He divided nature into the kingdoms of animals, plants, and minerals, and subdivided the animal kingdom into six classes. The classes were in turn divided into orders, orders in genera, and genera into species. Members of a species, Linnaeus found, were able to interbreed and produce other members of that species.

Darwin used classification to trace the origin of species. He grouped species according to common characteristics and then traced these characteristics back to the ancestor. For example, when he studied domestic pigeons, he noted the many differences between breeds. However, Darwin found that, despite these many differences, all domestic pigeons descended from the rock pigeon (*Columbia livia*). He supported his theory by tracing the common characteristic that pigeons inherited from the rock pigeon—coloring:

> Some facts in regard to the colouring of pigeons well deserve consideration. The rock pigeon is of a slaty-blue, with white loins, but the Indian sub-species, C, intermedia of Strickland, has this part bluish. The tail has a terminal dark bar, with the outer feathers externally edged at the base with white. The wings have two black bars. Some semi-domestic breeds, and some truly wild breeds, have, besides the two black bars, the wings chequered with black. These several marks do not occur together in any other species of the whole family. Now, in every one of the domestic breeds, taking thoroughly well-bred birds, all the above marks, even to the white edging of the outer tail feathers, sometimes concur perfectly developed. Moreover, when birds belonging to two or more distinct breeds are crossed, none of which are blue or have any of the above-specified marks, the mongrel offspring are very apt suddenly to acquire these characters. To give one instance out of several which I have observed:—I crossed some white fantails, which breed very true, with some black barbs—and it so happens that blue varieties of barbs are so rare that I never heard of an instance in England; and the mongrels were black, brown, and mottled. I also

crossed a barb with a spot, which is a white bird with a red tail and red spot on the forehead, and which notoriously breeds very true; the mongrels were dusky and mottled. I then crossed one of the mongrel barb fantails with a mongrel barb spot, and they produced a bird of as beautiful a blue colour, with the white loins, double black wing bar, and barred and white-edged tail feathers, as any wild rock pigeon! We can understand these facts, on the well-known principle of reversion to ancestral characters, if all the domestic breeds are descended from the rock pigeon. . . .

Darwin found the "origin" or ancestor of domestic pigeons by tracing the dominant characteristic of domestic pigeons back to the ancestor who bore that characteristic. Classification then is not only a way of grouping subjects, but also a means of analyzing them.

## The Bases of Classification

When we classify subjects, we group them according to a meaningful characteristic that all subjects in one group share. This basis can be a characteristic of structure, such as a classification of books according to whether they have a soft or hard binding. The basis can be a characteristic of use or function, such as a classification of kitchen utensils according to whether we use them to prepare or to consume food. The material or substance of our subjects can be a basis, such as a classification of carpets according to fiber content, or our kitchen utensils according to whether they are made of wood, metal, or plastic. Finally, we can classify according to cause or origin, such as in a classification of diseases. Whatever basis we choose, it must be appropriate to our audience and purpose. For example, we might classify bicycles according to gear complexity for a bicycle repair man and according to price or size for a potential customer.

## Partition

Partition is a special type of classification. While we classify many subjects, we *partition* one subject, a subject usually seen as a whole. Partition is similar to device description, but instead of dividing the device into "major parts," we group the parts according to a basis. For example, if we repair typewriters, we might partition a typewriter according to parts made of rubber, plastic, or metal before analyzing durability or ordering replacement parts. Although we could classify many typewriters according to size or price, we could partition *one* typewriter according to material or parts. As in classification, we can partition by material, structure, function, and cause.

## Classification and Partition Based on Structure

One of the most common and useful ways to classify or partition is on the basis of structure. Since we help an audience visualize a subject by describing its parts, we can group many subjects according to the most important part.

For example, in the following classification, the writer initially divides single-exposure 35-mm cameras according to structure, or specifically according to their focusing mechanism. The names of the main groups are: rangefinder and single-lens-reflex cameras. The writer discusses one group, rangefinders, completely before going on to single-lens-reflex cameras. He then subdivides single-lens-reflex cameras according to their metering systems: manual or semiautomatic, again a structural basis but one dependent on more specific parts. After discussing manual metering systems thoroughly, the writer subdivides semiautomatic systems according to whether the aperture setting is set manually and the shutter speed set automatically or vice versa: aperture-priority (manually) or shutter-priority (manually). Notice that each time the writer subcategorizes, he makes clear what basis he is using. He develops the subdivisions in his classification by using device description, definition, and causal analysis:

Many people while contemplating the purchase of a new camera these days are baffled and confused by the overwhelming number of makes and models on the market. This is especially true of single-exposure 35-mm cameras which have the most diverse selection available. These cameras can be most easily classified according to their focusing mechanisms. On this basis there are two main types: *rangefinder* and *single-lens-reflex.*

The *rangefinder* camera is focused using an optical rangefinder which is attached to the lens barrel. The rangefinder has two openings toward the top of the camera which allow light to enter into a series of mirrors and lenses. These mirrors and lenses superimpose the two images on a ground glass screen in the viewfinder. The camera is focused by moving the lens barrel back and forth which in turn moves the mirrors and lenses in the viewfinder. The camera is in focus when the two images in the viewfinder coincide exactly.

The *single-lens-reflex* camera (SLR) is focused without using any external optical systems. Instead the light entering the lens barrel is directed onto a prism using a mirror in back of the lens barrel (the mirror moves out of the way when the picture is being taken). The prism focuses the light entering it from the mirror onto a ground glass screen in the viewfinder.

Most SLR's have what is called "through-the-lens metering," or, in other words, a system that determines the proper exposure by measuring the light entering through the lens. This system eliminates the need for bulky light-meters that are needed for many rangefinder cameras. Thus SLR's can easily be classified further according to the type of metering system. On this basis there are two main types: *manual* and *semiautomatic.*

The *manual* system determines proper exposure by first converting the light entering the lens into an electrical voltage using a cadmium-sulfide (CdS) cell. The output voltage of this type of cell is proportional to the amount of light striking the cell. This output voltage is analyzed by an electronic device that compares this signal to other signals from the film-speed dial and the position of the aperture and shutter-speed rings. The result is an output voltage of a calculated value which is sent to a galvanometer located to one side of the viewfinder. Correct exposure is determined by rotating one or both of the aperture and shutter-speed rings until the galvanometer needle comes to rest somewhere between the two marks located on the same side of the viewfinder as mentioned above. Notice that both the shutter speed and aperture are set by hand.

The *semiautomatic* system on the other hand will set either the shutter speed or the aperture automatically. There are two types of semiautomatic systems: *aperture-priority* and *shutter-priority*. The *aperture-priority* system is so named because the aperture setting is set manually and the shutter speed set automatically. In this system the electronic circuits use the inputs from the CdS cell, the film-speed dial, and the aperture ring to determine the proper shutter speed. The *shutter-priority* system works in exactly the opposite manner. In this case the shutter speed is set manually and the aperture set automatically. The aperture is determined from the inputs from the CdS cell, film-speed dial, and shutter-speed ring.

While it is possible to get through-the-lens metering in a rangefinder camera, it is by far the exception rather than the rule. For this reason alone, SLR's are probably the best choice for most amateur photographers. In addition, the through-the-lens focusing found on SLR's allows the possibility of using other than the standard lens, for example, telephoto and wide-angles lenses. This system is available on most SLR's on the market. Of course there are also other considerations such as weight, quality of lens, and price, but this general outline should help potential buyers determine what the best camera is for them.

*Lawrence Ching*

We usually classify or partition according to structure for technical audiences, those audiences interested in the design of our subject.

## Classification and Partition Based on Function

We can base less complex classifications and partitions for audiences with less technical background and experience on the function of the subjects or parts. Since function is the effect or result of structure, or since the makeup of the subject determines what it can do, we use function for our basis when we address an audience more interested in the *what* than in the *why* of our subject.

For example, in the following partition on the human eye, the writer addresses a nontechnical audience, one interested in *what* the parts of the eye do. He partitions or groups the nine parts of the eye according to one of five basic functions:

The human eye is composed of nine main parts: the eyebrow, the eyelid, the eyelash, the iris, the pupil, the cornea, the retina, the optic nerve, and the eyeball muscles. These parts can all be categorized according to their functions. There are five functions: protection, regulation of amount of light entering the eye, transportation of impulses, focusing, and movement.

The eyebrow, the eyelid, and the eyelash all protect the eye. The eyebrow and eyelash both catch tiny dust and dirt particles before they can reach the eyeball. The eyelid protects the eye from larger objects and from wind and water.

The iris and pupil function as light regulators. When the iris dilates, the pupil, which is an opening in the eyeball, becomes larger. More light is allowed to enter. When the iris contracts, the pupil becomes smaller, and less light is allowed to enter the eye.

The retina and optic nerves are used to transport the impulses from the eye to the brain.

The cornea, which is the thin, clear part on the front of the eye, acts as a lens to focus the light on the retina. This makes the vision clear.

The eyeball muscles are used to enable the eyeball to move about within the socket.

*David Duffy*

This partition could introduce a detailed description of the eye or instructions on how to care for the eye.

When we classify or partition on the basis of function, our discussion is less complex and less technical than classification or partition based on structure. Our communication would probably be directed upward to managers interested in the effect of our subjects or downward to potential customers interested in the use of our subjects.

## Classification and Partition Based on Material

We can also classify subjects and partition parts of subjects according to the material or substances which make up the subjects or parts. Although material is really a part of structure, sometimes the materials that make up the subjects or parts of subjects tell an audience about durability, replacement, and function.

For example, in the following classification, the writer groups cleansing agents according to the *materials* used in manufacturing them. She then subdivides the main class of synthetic detergents according to the *materials* used to achieve certain results:

Most cleansing agents are referred to inappropriately as soaps. The materials constructing an actual soap are natural products, while highly advanced chemical technology is used to produce synthetic detergents, or syndets. There has been a sharp increase in the use of these syndets in America, the nation with one of the highest rates of per capita consumption of cleansing agents. The majority of soap or soaplike products used are synthetic detergents. These two groups of cleansers, detergents and soaps, can be classified on the basis of the materials used in their manufacture.

Soap production is characterized by the simple chemical reaction between its natural ingredients. The women of the pioneer household were delegated the task of stirring and preparing the soap for their households. The basic components of a natural soap are a water insoluble fatty acid, and an organic base or alkali metal. The result of the manufacture is a partially water soluble solution.The disadvantages of this natural process are apparent in the nature of its chemical reaction with calcium ions often found in water.

The once inferior syndet has sharply displaced soap as leader of the cleansers. Initially a German substitute for soap, detergent was developed to avoid use of precious fats during World War II. Technology has advanced synthetic detergents to such a high level that they are the ideal cleansers. The reasons for the increased use of detergent can be classified by its solutions to problems of soap use. The two major advantages of detergents are their failure to form a precipitate which is deposited on the item being cleansed and their prevention of lack-lustre hair.

The calcium found in water reacts with soap ingredients to form spots on glassware and brown discoloration on laundry. However, detergents let the water roll off glassware without a trace as the residues join with the water molecule. The brown spots on fabrics are avoided because the detergents do not react with the calcium found in hard water to form a discoloring acid.

Before the advent of synthetic detergents it was necessary to use harsh soap to wash hair. This was detrimental to the general health and appearance of the hair as dull films were left and natural oils removed. Experiments with shampoo production have resulted in two major methods of manufacture. These methods can be classified by the chemicals used in production which have different cleansing effects. On this basis there are two chemical processes used for production of shampoos: sulfonate acids and alkyl sulfates.

The sulfonate acid process was the first developed and is rarely used in modern production. The process uses sulfur, an excellent hair cleanser. Unfortunately the unbuffered cleanser is so effective that it strips the hair of all its natural oils. This produces hair that is brittle and unhealthy. Use of this product necessitates various creme rinses and conditioners.

The simple addition of one atom of oxygen to form sulfates instead of sulfonates produces a very effective hair cleanser. The sulfates remove excess oil, but leave enough of the natural oils for a healthy head of hair. Aiding the sulfates are a variety of auxiliary ingredients, which vary according to manufacturer, and organic amines. The organic amines are used as foam stabilizers to produce a full head of lather, while the auxiliary ingredients give the scent and other characteristics for a particular brand of shampoo.

Instead of reaching for a bar of lye soap for hair use, the average American is now reaching for a bottle of one of hundreds of name brands of shampoo. These shampoos vary little in basic cleansing ability because most use the same sulfate base. This cleanser preserves and strengthens the hair as it is cleaned. It has been proven in tests that the most effective results are found when these sulfate-base shampoos are used. Once again, the chemist has aided the beautifying process of the American woman: the billion dollar shampoo market is a strong indication of this.

*Kim Dellas*

Although a nontechnical audience could understand this classification, the writer explains the differences between the main categories of cleansing agents according to their chemical makeup. However, instead of giving the complete chemical structure of the agents, she describes only the most important chemical ingredient. If a classification or partition according to structure would be too complex for an audience, and a classification or partition based on function not detailed enough, we can use material—the main substance—of our subject as our basis for grouping.

# Classification and Partition Based on Cause

If our purpose is to discuss how to control or change a great many subjects, we can classify or partition according to cause. For example, in the following classification, David G. Lygre groups all genetic diseases according to cause: too few or too many chromosomes, several defective genes, or a single defective gene. Because he addresses a nontechnical audience not interested in or able to understand the complex structure and function of chromosomes, Lygre classifies the diseases according to their basic cause:

> We can classify the two thousand or so genetic diseases into three general groups. In one category (gross chromosome abnormalities) are disorders where patients have too many or too few chromosomes in their cells, or parts of their chromosomes are misplaced. For example, people with the most common form of Down's syndrome have an extra chromosome number 21—three instead of the usual pair. Others may have an unusual number of sex chromosomes, such as Klinefelter's syndrome (XXY), Turner's syndrome (X), XXX, or XYY. By inspecting a karotype, a visual display of the patient's chromosomes arranged in a standard pattern, a physician can readily diagnose such abnormalities. But no cure is in sight for these disorders. All the doctor can do is treat symptoms.
>
> In the second category (polygenic disorders) are people whose chromosomes look normal but actually carry several defective genes. Environmental factors may also be important in some of these disabilities. In this potpourri are congenital heart disease, cleft palate, club feet, spina bifida, anencephaly, diabetes, and schizophrenia. Physicians can provide effective therapy for some of these disorders.
>
> More than one hundred diseases belong to the third group—single gene defects. The list includes phenylketonuria (PKU), sickle-cell anemia, galactosemia, Tay-Sachs disease, hemophilia, cystic fibrosis, Duchenne muscular dystrophy, and Huntington's chorea. Scientists have pinpointed the problem and devised specific therapy for a few (PKU, galactosemia, hemophilia). But for others, they either have not unraveled the chemical defect (Huntington's chorea, Duchenne muscular dystrophy), or they know the problem but not a solution (sickle-cell anemia, Tay-Sachs disease).

Once Lygre has classified disease according to cause, he discusses treatment. Grouping diseases according to cause helps Lygre's audience understand a great many diseases that might at first seem unrelated.

# Pattern and Style of Classification and Partition

Whether we classify or partition according to structure, function, material, or cause, we should divide and subdivide as completely as possible with no overlaps between groups. We must choose a basis that is meaningful to our audience and relevant to our subject. Generally we should use the same basis throughout the classification or partition. However, if we refine our basis to subdivide the main groups, we should always announce the refinement to the audience. For example, in the classification of cleansing agents, the main division was based on materials used in manufacture and the subdivision based on materials chosen to produce special effects. We always state the specific basis for each subdivision we make.

Before we begin our classification or partition, we must identify our subjects. If the subjects are new or unfamiliar to our audience, we might open our discussion with a definition. We should also specify our basis for initial grouping and the names of the groups. Then we discuss each category or group in the body of the classification or partition before going on to the next.

For example, we discuss completely rangefinders and any subcategories before discussing single-lens-reflex cameras. We can begin our discussion with the simplest or the most complex main categories depending on our audience's needs. Although the writer of the single-exposure 35-mm camera classification described rangefinders first, he could have discussed single-lens-reflex cameras and all subdivisions before discussing rangefinders.

## Charting to Check and Clarify Classification and Partition

To check the completeness of our classification or partition and to eliminate any overlap between categories, we can outline or chart our division and subdivisions. Such an outline or chart sometimes accompanies a classification or partition as a visual aid. If we chart the classification of single-exposure 35-mm cameras, we have the following:

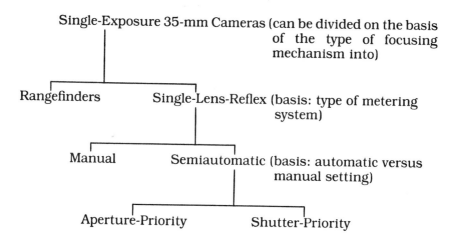

Single-Exposure 35-mm Cameras (can be divided on the basis of the type of focusing mechanism into)

Rangefinders     Single-Lens-Reflex (basis: type of metering system)

Manual     Semiautomatic (basis: automatic versus manual setting)

Aperture-Priority     Shutter-Priority

Charting or outlining a classification or partition helps us make sure that we have subdivided as completely as possible and that we have made clear our basis for each category. When we develop our classification or partition, we discuss the categories as charted vertically rather than horizontally.

An outline at the beginning of a classification or partition acts much like a table of contents. This outline helps an audience anticipate the subdivisions and overall structure of the upcoming classification or partition. The writer of the discussion of single-exposure 35-mm cameras could have outlined his classification as follows:

Single-Exposure 35-mm Cameras

 I. Rangefinders
 II. Single-Lens-Reflex
    A. Manual
    B. Semiautomatic
       1. Aperture-priority
       2. Shutter-priority

We can use either an outline or chart to check the clarity and completeness of our classification or partition or to support the writing as a visual aid.

# Classification and Partition in Technical Communication

Let us look now at some of the specific uses of classification and partition in professional communication.

## Classification and Partition as Introduction

An introduction usually establishes a common background between the writer and the audience as well as gives a preview of what is to come in the communication. One way to do both tasks is to classify or partition the subjects to be discussed. The classification or partition can indicate both the basis for and the organization of the discussion. When using classification or partition to introduce a communication, we can not only group our subjects according to a significant basis, but we can also present a thorough "table of contents":

For example, in the following partition of computer systems based on function, the writer introduces the activities at one Computer Science Center. The partition helps the audience understand and anticipate the discussion that follows. Notice that the partition begins with a definition:

A computer is a "tool" that processes information. Input to this system is an application program and its associated data. Output from the system is the processed data in the form of calculated results or new data bases. Computer systems are composed of two basic parts: hardware and software. Computer hardware consists of an integrated set of electronic and mechanical components that perform operations specified by the software. There are three general classes of software:

Application program, which is the user's program, usually written in a higher level computer language, to accomplish a specific task;

System support software, which is the system software that translates the application program or command language into instructions acceptable to the hardware;

Operating system software, which is the system software that controls and allocates the hardware and software resources, thus allowing the application program and system support software to perform their functions.

The writer went on to discuss each category in the order presented in the partition.

Thus we can use a classification or partition to introduce a communication that deals with a wide range of related subjects and to identify the basis for their relationship. We can also use the classification or partition to preview the organizational pattern of the communication.

## Classification and Partition of On-the-Job Procedures

Often classification or partition can help an audience understand new procedures or make better decisions when performing familiar tasks. We classify or partition duties, or equipment used in these duties, so that employees, users, or technicians can judge how to complete a task. Although in these cases we would be writing downward-directed communication, we are classifying or partitioning rather than instructing. We explain into what categories a job or a tool would fall rather than specifying step-by-step performance.

In the next example the writer explains computer procedure by partition. The audience is composed of users who know how to operate a computer but who need to understand the different computers in the whole system or network. The writer partitions the network on the basis of performance or function:

The fact that the Integrated Computer Network (ICN) is divided into three computing partitions can be ignored if you use the same computer all the time. However, if you use different computers (at least, if the computers you use are in different computing partitions), you must know at least the basics of partitioning.

The three partitions are called the Secure (Red), Administrative (Blue), and Open (Green). The machines in the Red partitions (LTSS Machines, R, S, and T and DEMOS Machine V) and those in the Green partition (LTSS Machine U and NOS Machine M) are connected to the Common File System (CFS); the machines in the Blue partition (NOS Machines L and N) are not connected.

The restrictions on the operations that can be performed from machines in the various partitions are as follows:

1. Classified computing can be done only in the Red partition.
2. Files saved in CFS from worker computers in the Red partition *cannot* be retrieved by worker computers in the Green partition. However, files saved in CFS from worker computers in the Green

partition *can* be retrieved from CFS by worker computers in the Red partition.

3. Files saved from a machine in one partition cannot be replaced from a machine in another partition.

4. Files saved in CFS from a worker computer in the Red partition *cannot* be deleted from a worker computer in the Green partition. However, files saved from a worker computer in the Green partition *can* be deleted from a worker computer in the Red partition.

Again, although the audiences of the partition and the classification above know their jobs well, the partition and classification help them make better decisions when performing those jobs. We can use classification and partition in downward-directed communications to explain on-the-job procedures to technical or semitechnical audiences.

## Classification and Partition to Recommend Alternatives

We can also use classification and partition to analyze choices or alternatives and to recommend one of these to an audience. If we have a great many possible solutions to a problem or alternative actions or approaches, we can group our choices according to common characteristics. This grouping enables us to analyze logically our choices and present the best recommendation to our audience. We can classify not only our choices but also our criteria for making a choice. However, whether we classify alternatives or criteria, we identify the basis for grouping as the most important characteristic of either alternatives or criteria. Our analysis and recommendations are then clear and logical.

For example, in a communication on how to select computer services, the writer classified performance criteria according to use or value. The two main groups of performance criteria are measurable and nonmeasurable characteristics. Each of these groups are subdivided into mandatory and desirable criteria. Classifying the criteria for selecting a computer helps the audience choose between computer systems. Each system meets some, all, or none of the criteria identified as being most important to the potential buyer and user:

> In choosing the best computer service from several alternative computer services, performance criteria must be defined which describe what is meant by best. These criteria can be divided into those for which no empirical measurement is necessary and those whose values are derived from actual system measurements. For example, evaluating system documentation and amount of main memory does not require measurement-collecting activity, whereas evaluating system turnaround time and response time clearly requires measurement.
>
> Performance criteria can further be divided into those which are mandatory and those which are desirable. A mandatory criterion is defined to be any performance requirement which must be satisfied by the computer services being considered for selection. Desirable criteria, on the other hand, are those which are not absolute require-

ments for system acceptance, but which would make the implementation of the purchaser's work easier. Therefore, if a given computer service does not include some desirable features, it would continue to be considered for selection, but the lack of each desirable feature would perhaps invoke some penalty on the respective computer service. . . .

Based on the two characteristics described above, performance criteria can be classified as: Mandatory Nonmeasurable (MN), Mandatory Measurable (MM), Desirable Nonmeasurable (DN), and Desirable Measurable (DM). Examples of each class of criteria are provided in Table 7-1.

**TABLE 7-1. Examples of Performance Criteria**

| Type | Example |
| --- | --- |
| Mandatory Nonmeasurable | 1. The system must be fully delivered and operational no later than September 1, 1979. |
| | 2. Time-sharing service must include FORTRAN, Basic, Lisp, SNOBOL and editing facilities. |
| Mandatory Measurable | 1. The mean-time-to-failure for a specific one-month period must be greater than 4 hours. |
| | 2. 95% of all trivial command response times must be less than 1 second. |
| Desirable Nonmeasurable | 1. It is desirable that the system include Pascal and COBOL facilities. |
| | 2. It is desired that the system provide a text-editing capability. |
| Desirable Measurable | 1. It is desired that the system provide a mean turnaround time for the benchmark run of 5 minutes or less. |
| | 2. It is desired that 95% of all trivial command response times be 0.5 seconds or less. |

Using classification or partition to analyze and select recommendations draws a logical frame around our discussion. In selecting a basis for classification or partition, we must analyze and then choose the most important characteristic of our alternatives. This basis, and the classification or partition organized according to that basis, help us recommend the alternative which best meets the audience's needs.

# Summary

When we classify, we group many subjects according to a common characteristic. This characteristic forms the basis for our classification. We can also partition one unit or subject by grouping its parts according to a basis. The most useful and common basis for both classification and partition is structure. However, if our audience cannot understand the complexities of structure, we can use the result of structure, function, as a basis. In some cases, we can also classify or partition according to cause or material.

Whether we classify many subjects or partition one subject, our categories and subcategories should not overlap, and if possible, we should use one basis throughout. If we redefine a basis to sub-categorize, we should state the new basis and justify our change to our audience. A chart or outline clarifies any classification or partition for both writer and audience.

In technical communication, we often introduce our subjects and preview our discussion by classification or partition. We also can use classification or partition to help experienced audiences make better decisions on the job. Using classification and partition to discuss alternative recommendations helps us analyze logically our choices or our criteria for these choices.

# Readings
# to Analyze and Discuss

## I

Basically, there are three types of communication problems in business: those related to skill and knowledge, motivational problems, and problems of information. . . .

Problems in the skill-knowledge area usually involve situations where an employee lacks the understanding necessary to perform a job. People have to be trained to make products. People have to be instructed in safety procedures. People have to be coached in selling the product. The idea is this: People who know more about their jobs are more effective. That's a good investment.

In the motivational area, the problem is that even though some employees can do a job, they may not *want* to do it. Most companies recognize that people who want to work are more productive and will work harder toward a solution to the problem. Motivational problems can also involve a company's prospective clients. The solution in this case can usually be found in the area of more effective advertising and sales promotion.

In the informational area, the problem is what most people would refer to as "public relations." Corporate and product visibility is very important to most companies, since exposure and goodwill help sell products. A company that perceives a need for solving informational problems will invest in the solution that best reaches its audience.

## Questions for Discussion

1. Is the classification above based on structure, function, material, or cause? Why did the writer choose this basis? What are the audience and purpose of the classification?

2. Can you think of an example to fit into each category of problems? Could the categories be subdivided further? How?

3. Can you apply these three categories to a business with which you are familiar?

## II
## He's Not My Type

For purposes of transfusion there are four types of human blood [O, A, B, AB], but the number is greater from the genetic viewpoint. Every person inherits two genes governing the particular blood groups I have been discussing, one from his mother and one from his father. Each gene can bring about the production of A, or B, or of neither, so that the genes are spoken of as belonging to the A, B, O group.

You can inherit any of six possible combinations then: OO, AO, AA, BO, BB, AB. When you possess the AO combination, the one A gene brings about the production of A corpuscles just as well as two A genes would. You are of blood type A, then, whether your combination is AA or AO. By similar reasoning, you are of blood type B, whether your combination is BB or BO. Your gene combination is your "genotype" and what you actually appear to be by test is your "phenotype." In other words, the six possible genotypes work out to four phenotypes.

But, you may ask, what does it matter whether you are AO or AA? Your blood reacts equally in either case, so why make a point of it? As far as transfusion goes, to be sure, the difference is negligible. But consider—

If two AA individuals marry, each can contribute only A genes to their offspring. All their offspring *must* be of blood type A. On the other hand, if two AO individuals marry, then it is possible that each will contribute an O gene to a particular offspring, which will then be OO and will test out as blood type O.

In other words, if two people, both of blood type A, marry, it is possible for an offspring to be of blood type O, without any hanky-panky having been involved. The existence of the AO genotype as opposed to the AA genotype is thus very important in paternity suits.

It was eventually found that there were two kinds of A corpuscles, one that reacted strongly with anti-A, and one that reacted weakly. The former was called $A_1$ and the latter $A_2$. This difference is of little importance in transfusion, but is, again, significant in paternity suits, since, for example, two $A_1$ parents cannot have an $A_2$ child, and vice versa.

## Questions for Discussion

1. What is the basis for the classification of blood types? Why does the writer classify the genotypes rather than the phenotypes? Can you define genotype and phenotype in a formal definition?

2. Chart the classification of genotypes.

3. What are the audience and purpose of the classification? Give evidence to support your answer.

4. Place yourself in the role of a defense attorney in a paternity suit. The infant's genotype is $A_1B$, the mother's is $A_2O$, and your client's is $A_2B$. Can you defend your client? How? If the infant's genotype is OO, the mother's $A_1O$, and your client's BO?

# Exercises and Assignments

1. **Problem:** As we have studied, the basis we choose when classifying or partitioning a subject depends on our audience and purpose. In most cases, we can classify or partition any subject according to structure, function, material, or cause. Usually we can classify the types of our subject and partition the parts of any of these types.

Assignment:

Analyze the subjects below and specify what basis you would use to classify or to partition each subject for each audience listed. If possible, choose a basis for both classification and partition.

| Subject · | Audience |
|---|---|
| Trees | Botanists |
| | Lumbermen |
| | Owners of a wood stove |
| Automobiles | New car salespersons |
| | Used car salespersons |
| | Insurance companies |
| | Teenagers |
| | Senior citizens |
| Dogs | American Kennel Club |
| | Dogcatchers |
| | Pet store owners |
| | Mail carriers |
| Tennis rackets | Re-stringers |
| | Sports equipment store managers |
| | Pro tennis players |
| Hammers | Carpenters |
| | Hardware store owners |
| | Homeowners |
| | Metalworkers |
| Doctors | American Medical Association |
| | Patients |
| | Nurses |
| | Doctors |

2. **Problem:** Each sketch in Figure 7-1 represents one of four main types of bridges. Study the sketches carefully to decide what characteristic distinguishes these bridges and determine whether that characteristic is based on structure, function, material, or cause.

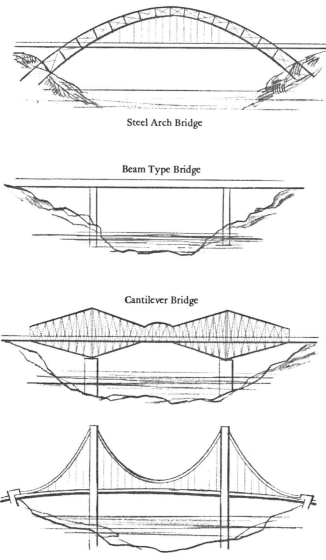

Steel Arch Bridge

Beam Type Bridge

Cantilever Bridge

Suspension Bridge

*Figure 7-1.* From *How Does It Work?* by Richard M. Koff and illustrations by Richard E. Rooman. Copyright © 1961 by Richard M. Koff. Reproduced by permission of Doubleday & Company, Inc.

## Assignment:

Write a classification of bridges based on the sketches in Figure 7-1. State the basis for the classification and devote a paragraph to each type of bridge. Choose and state a specific audience and purpose.

**3. Problem:** In Figure 7-2, the two main types of fishing reels are depicted: the rotating spool and the spinning reel. The open-face spinning reel and the enclosed-spoon spinning reel are subcategories of the spinning reel shown in the cut-away sketch.

**Assignment:**

Using the characteristics you can observe in Figure 7-2, write a classification of fishing reels. State the basis for your classification, devote a paragraph to each main category and subcategory, and choose and specify an audience and purpose.

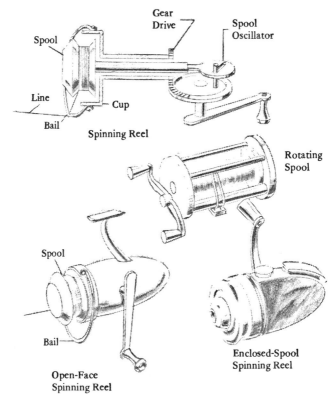

*Figure 7-2.* From *How Does It Work?* by Richard M. Koff and illustrations by Richard E. Rooman. Copyright © 1961 by Richard M. Koff. Reproduced by permission of Doubleday & Company, Inc.

**4. Problem:** You have been asked by your supervisor to conduct a workshop on communication for your employees. You would like to give introductory comments on the theory and types of communication before you give concrete suggestions on how to improve communication within your department. You have placed three types of communication and their characteristics in Table 7-2, but need to add some specific examples of each. You also have to decide whether the three types of communication you have listed—intrapersonal communication, interpersonal communication, and mass media—are main categories or whether they are subcategories of major groups reflected in the characteristics in the table. If the characteristics in the table reflect a major division, you need to choose which characteristic is most important or distinctive. You may add details about the situation if you wish.

**Assignment:**

Write a classification of communication using the characteristics and types identified in Table 7-2. Add examples which will directly concern your audience to illustrate each type of communication. Decide whether the main divisions appear at the top of the table or are reflected in the body. Support your classification with a chart.

**TABLE 7-2  Communication**

|  | Intrapersonal | Interpersonal | Mass Media |
|---|---|---|---|
| Situation | Internal thoughts of one person | Face to face communication, two or more people | Message sent through channel to unknown mass audience |
| Communicator of message | Personal and known | Personal and known | Impersonal |
| Feedback | Immediate and direct | Immediate | Delayed |
| Type of message | Tailored to audience | Tailored to audience | "To whom it may concern" message |

**5. Problem:**  Partition is especially useful in categorizing the duties and responsibilities of individuals in organizations.

**Assignment:**

Partition a local or campus organization in which you have been active. Divide and subdivide the groups of people involved in the organization according to their duties and responsibilities or some other characteristic. Your basis for partition will depend on the audience and purpose you specify.

**6. Problem:**  As we studied, in classification we place many subjects into groups based on a characteristic of structure, function, material, or cause, while in partition we group the parts of one subject according to one of these bases. To check for clarity and completeness we can chart our classification or partition. We can use classification to introduce subjects, recommend subjects, or clarify on-the-job duties.

### Assignment:

1. Choose a subject with which you are familiar.
2. Place yourself in a hypothetical situation and designate an audience and purpose.
3. Write a classification of your subject.
4. Use a chart to clarify your classification for your audience.

### 7. Assignment:

Using the same subject you chose for Assignment 6, partition one type of your subject. For example, if you were to classify types of bicycles, you could then partition one of these types by grouping its parts according to a basis. If you were to classify types of trees, you could then partition one tree by grouping its parts according to a basis. You may refine or change your audience and purpose of Assignment 6 for your partition.

# PART FOUR

## Presenting Subjects in Motion or in Use to an Audience

# Narrative:
## Selecting
## and Sequencing
## Events in Time

*Choosing and expressing point of view in narrative.*
*Selecting steps or details to recreate an event or*
  *occurrence.*
*Sequencing these steps chronologically or developmentally*
  *(as they actually happened, or reordered to lead to a*
  *climax).*
*Choosing a starting and ending point in narrative.*
*Indicating causality in narrative.*
*Using narrative to introduce or organize technical*
  *communications.*
*Reading and analyzing narratives written by others.*

We now know how to identify and describe one subject and how to present many subjects to an audience. When using these rhetorical patterns, we have often treated subjects as if they were separated from their changing environments and frozen in time. Especially when we define, describe, compare and contrast, and classify subjects, we present a subject in great detail to our audience, so highlighting that subject for a moment that everything surrounding it fades away.

However, after we explain the parts, species and classes, or main characteristics of our subjects, we might want to describe them in motion. We might enhance our description, definition, or classification by showing the action, operation, or development of our subjects. Machines operate; species grow and change; scientists experiment and discover; human culture progresses and changes; technical writers review previous investigations—in technical communication we often describe or recreate movement, select and sequence events in time. The basic rhetorical pattern we use to present action, events in time and motion, is narrative.

Although often we associate narrative with fiction, technical communicators often recreate events, developments, and experiments for their audiences. Narrative recreates an event or gives a *sense* of it by helping the audience visualize that event. When we tell of any event, we select what we think are the most important aspects of that event. We place our subject in time and relate the beginning, middle, and end of it. We tell of the event ourselves, choose a narrator to tell it for us, or present it so objectively that we hope the audience forgets someone must tell the tale. We may reorder what happened so that we lead up to a climax or tell of the event chronologically in the actual sequence in which it occurred. We may be involved in the event ourselves, observe it from the sidelines, or retell someone else's experiences. In any case, no one reproduces life exactly for an audience. Even in our best efforts to duplicate a true event in words, what we select and how we order our selections reflect our own sense of what is important. By narrative, scientists and technicians recreate in words the *essence* of occurrences and events for their audiences.

For example, in 1901 Maurice Maeterlinck, a natural historian, studied the life of the bee. Maeterlinck, in the following narrative passage, recreates for his audience the queen bee's destruction of the young queens who might form a second swarm:

> . . . . Our young queen hastens toward the large cradles, urged on by her great desire, and the guard make way before her. Listening only to her furious jealousy, she will fling herself onto the first cell she comes across, madly strip off the wax with her teeth and claws, tear away the cocoon that carpets the cell, and divest the sleeping princess of every covering. If her rival should be already recognizable, the queen will turn so that her sting may enter the capsule, and will frantically stab it with her venomous weapon until the victim perishes. She then becomes calmer, appeased by the death that puts a term to the hatred of every creature; she withdraws her sting, hurries to the adjoining cell, attacks it and opens it, passing it by should she find in it only an

imperfect larva or nymph; nor does she pause till, at last, exhausted and breathless, her claws and teeth glide harmless over the waxen walls.

The bees that surround her have calmly watched her fury, have stood by, inactive, moving only to leave her path clear; but no sooner has a cell been pierced and laid waste than they eagerly flock to it, drag out the corpse of the ravished nymph or the still living larva, and thrust it forth from the hive, thereupon gorging themselves with the precious royal jelly that adheres to the sides of the cell. And finally, when the queen has become too weak to persist in her passion, they will themselves complete the massacre of the innocents; and the sovereign race, and their dwellings, will all disappear. . . .

Although Maeterlinck narrates an event that happens again and again in the lifetime of a hive, he individualizes the bees for his audience. He describes the queen's fury, the workers' hostility, and the young queens' violent death. As if the queen were a human character, we know her thoughts as well as actions. The narrative has a beginning, a climax, and an end. Maeterlinck has reproduced as closely as possible the essence of the hive by encouraging his audience to imagine all the emotions and actions of the queen and workers.

In this chapter we study the process of selecting and sequencing events in time. This will serve as the foundation for the next three chapters, which deal with process description, a type of narrative we use in technical communication to relate the operation of a device, and procedure, instructions, and specifications, other technical narratives that we use to handle an operation, operate or fix a device, and meet the criteria of a future project.

# Point of View

When we recreate an event for an audience, we ask that audience to "participate" in our narrative by visualizing what happened, by forgetting time now and imagining time then, by relating to the emotions of the characters involved, and by anticipating the climax or ending of our story. The audience also participates in narrative by relating to the teller of the narrative. Our choice of a narrator influences our audience's reaction to our narrative. If we narrate from our own point of view and appear as a participant in our story, the audience becomes involved with our thoughts as well as the actions we describe. If we choose another participant to narrate, the audience concentrates more on that person's thoughts and feelings than on ours. If we do not identify a narrator at all and describe only actions, not thoughts, the audience will react to the "what" of the narrative, a seemingly objective account of an event.

## First-Person Point of View

Basically, in narrative we choose between two main points of view: first person and third person. If we tell our own story, or create a narrator to tell it as if it happened to that narrator, we use the "I" or first-person point of view. This "I" can be a *participant* or an *observer.* If involved directly in the action, the narrator is a participant in the narrative; if the narrator looks on and relates someone else's actions, the narrator is an observer. For example, when we say, "I first measured a liter of the liquid and then added it to the prepared solution," we are writing from the first-person-participant point of view. When we say, "I watched Dr. Long first measure a liter of the liquid and then add it to the prepared solution," we are writing from the first-person-observer point of view. In the first case, our audience would be interested in our thoughts as well as actions. In the second case, the audience would be interested in our impressions of Dr. Long and his actions.

## Third-Person Point of View

When we relate a narrative that involves someone else entirely, in which we do not appear to the audience at all, we relate the narrative from the third-person point of view. Although we may have had to observe the action to be able to tell it, we do not remind the audience of our presence at the event or as we relate it. We also do not create a narrator who speaks directly to the audience as a recognizable character. For example, if we write "Dr. Long first measured a liter of the liquid and then added it to the prepared solution," we are using the third-person point of view.

There are three categories of third-person point of view based on whose thoughts and feelings we relate. While telling of the event, we might relate everyone's, one person's, or no one's thoughts. For example, if we write,

> First thing that morning, Chief Engineer Long wrote a memo announcing that the project had been dropped. He thought that if the news came from him rather than from the Project Director, Supervising Technician Jones would be less upset. Long delivered the memo personally. When Jones saw him, he sensed what was coming. Long tried but failed to mask his anger at the Project Director's decision and felt himself frowning. Jones leaned back in his chair and thought he must now hide his own disappointment.

we write from the *omniscient* point of view. Although we hope that the audience will forget the writer and concentrate on the characters, we also tell the audience what all the participants in the action are thinking—we are all knowing. On the other hand, if we write,

> When Supervising Technician Jones saw Chief Engineer Long typing first thing that morning, he sensed that Long had received word about the project. When Long came over a few moments later, Jones was shocked by the frown on Long's face. He didn't think that Long really cared about the project, but now it looked like he was very sympathetic. Or perhaps Long was just angry with him for suggesting the project in the first place. Jones leaned back in his chair and prepared to hide his own disappointment.

we are writing from a *limited* third-person point of view. We only know or relate the feelings of one character, Jones, the character with which we are most concerned.

Finally, if we write,

> Long, the Chief Engineer, typed a memo at 8:30 A.M. on Monday. The first sentence of the memo read, "Because of overall funding problems, Project Director Daniels discontinued Project A-168 until further notice." Long gave the memo to Jones, the Supervising Technician, at 9:00. Jones took the memo in his right hand and then leaned back in his chair. Frowning, Long stood in front of Jones's desk. After 2 or 3 minutes, Long returned to his own desk.

we relate only those actions that could be observed by anyone. We do not focus on any person's thoughts and feelings. The narrative seems to be an objective account of what happened. Of course, because someone always selects what to tell the audience, no narrative is completely objective. However, we use third-person-*objective* point of view when we want the audience to form its own conclusions about what the participants might have felt, when we have no idea what participants in the action were thinking, or when the participants' thoughts and feelings are not important to our narrative.

Thus first-person participant indicates that the narrator is an important character in the action; first-person observer that the narrator's impressions may be as important as the action; omniscient that the whole panorama of thought and action are necessary to visualize the event, third-person limited that the events will be seen through one person's eyes; and objective that although actions are depicted, individual impressions are not essential.

## Selection of Details in Narrative

When we recreate an incident by narrative, we select what to include and what to leave out. No one can tell the complete story, and words cannot imitate life exactly. We select detail according to audience and purpose of the narrative. For some audiences too many details distract; for other audiences too few details create a vague impression. Also we include in narrative only details that are relevant; yet in omitting details we do not want to distort the truth. We want to create a vivid and accurate impression of the event for the audience. Thus what to include in narrative is an essential decision.

For example, we can see how one incident can be narrated differently by two writers because they include different details. In the two selections that follow, the writers narrate a nuclear reactor shut-down at Three-Mile Island in Harrisburg, Pennsylvania, in 1979. The writers select detail for their narratives according to their audience and purpose. The Nuclear Regulatory Commission (NRC) presented the first to a senate subcommittee. Because the audience was unfamiliar with the operation of a reactor, the vocabulary could not be too technical. The NRC had concluded that the incident was caused by "operator actions" rather than "plant design" and recommended that other plants not be shut down. Notice what detail is included in the narrative and the point of view of this passage:

> ... At about 4 A.M. on March 28, the secondary cooling system suffered a malfunction. A malfunction in the main feedwater system caused the feedwater pumps to trip, which, in turn, caused the turbine-generator to trip. Since the steam generators were not removing heat, pressure in the reactor coolant system increased, and the pressurizer relief valve opened. The reactor scrammed.
>
> All this happened within the first 30 sec after initiation, and is considered a normal sequence. At this point, the three auxiliary feedwater pumps in the secondary system started up as designed, but were unable to deliver water to the system because of closed valves. After 8 min, valves were opened and flow was established. The pressurizer relief valve should have closed, but did not. This allowed the reactor coolant system pressure to fall. At a preset value of 1600 psi, the ECCS started as designed.
>
> At this point, an indication of a rapidly rising level in the pressurizer apparently misled the plant operators to shut off the ECCS flow. By now, the accident had been underway for about 12 min. ...

Because the purpose of this narrative was to convince the audience that the NRC's conclusions were correct, the narrative had to appear objective and unbiased. The point of view is third-person objective. The writer includes only enough detail to give the essence of the event and states the causes of each step in the event as fact.

Because the audience of the next narrative was made up of physicists, the vocabulary is more technical. A device description preceded the narrative. The writer gives a complete account of the operation of the reactor, the actions of the operators, and their probable thoughts as they acted. The writer includes so much detail because the purpose of the narrative is to speculate about all the causes of the incident. The "why" of the narrative unfolds as we read the "what." Notice the point of view used:

> The complex chain of events that made up the accident began when a condensate pump stopped, and the loss of suction in turn tripped the feedwater pumps. Among the various reasons suspected for the first pump failure are the spillage of ion-exchange resin from the demineralizer into the secondary flow, and the closing of an air-actuated valve because of moisture in the air line. (Just before the trouble began, an operator had been working on this portion of the feedwater system.)
>
> The loss of the feedwater pump resulted in the shutdown of the turbine. As pressure quickly rose in the primary loop, the reactor "scrammed." Nearly simultaneously, three auxiliary feedwater pumps began to operate. The water never entered the steam generators because the two valves had been left closed after a recent maintenance check.
>
> With the primary deprived of an adequate heat sink, its pressure rose above 2250 psi, at which point a relief valve atop the pressurizer opened. The pressure then dropped back through 2200 psi but, when the valve stuck open, the pressure continued to plummet. When the pressure declined to 1600 psi, the high-pressure injection system was automatically switched on, two minutes after the turbine tripped. One

high-pressure injection pump was turned off manually two minutes later, and a second ten minutes later. Both were restored again soon after that. The reason why the operator turned them off was probably his concern that a sensor in the pressurizer was registering a high level of water. . . .

. . . About one hour into the incident, an operator shut off the primary circulation pumps, apparently out of concern that they were badly vibrating, perhaps because of the bubbles in the system. Without circulation and apparently with little convection (a large temperature difference prevailed between inlet and outlet over this time), the core overheated. During the ensuing hours, the temperatures on some thermocouples on the fuel rods went above the computer-readout cutoff point of about 75°F. A large portion of the core was apparently uncovered after this point for an unknown period of time. . . .

Although the writer could only speculate about all the reasons for the operator's actions, the point of view seems almost omniscient. The operators' reactions are as important as a full, detailed account of the reactor response. By explaining the thinking processes that usually take place in the minds of the operators, the writer can speculate as to why the operators reacted as they did in this incident. Because the audience knows how a reactor works, the writer includes much more detail about each step in the incident. The final purpose of the narrative is to encourage this audience to study the incident, perhaps to prevent another.

What we include in narrative depends on our audience and purpose. Technical narrative always includes significant facts; the amount and type of detail that support these facts depend on our audience and purpose.

# Selection of Sequence in Narrative

Once we have chosen what details to include in our narrative, we have to decide in what order to present them. Do we start with the first thing that happened and follow straight through to the last? Do we state the final outcome and then explain how it came about? Do we show the importance or meaning of our narrative by interrupting it with anecdotes? The choices we make about how to order or sequence the stages of an incident are numerous, but they are choices that we make not only about narrative but also about many of the rhetorical patterns that we study in the rest of this book.

Basically we can sequence our narrative chronologically or developmentally. The chronological order is the natural sequence in time, which could be represented symbolically as A, B, C, D, E, F, etc. Detail, or stage A, would be the first thing that actually happened, B the second, C the third, and so on. The two narratives of the Three-Mile Island incident were basically ordered chronologically. Developmental order would be any variation on natural order: B, A, C, D, E, F; A, C, D, B, F, E; C, B, A, F, E, D; and so on. The possible patterns of developmental order are many. Why reorder the natural or chronological sequence of action? We might decide that stage B reveals important reasons for the incident and should come first, that stage C may catch the audience's attention and would be a good opener, or that stage D is most revealing and that we want to build suspense by leaving it for the last. Thus we can "reorder" time to clarify the narrative or make it more interesting.

In fact, to give meaning to what happens to us in life, we mentally reorder the natural sequence of events. For example, to understand why we lost a job, we might recreate certain scenes in our minds: what the boss said just before we lost the job, what the boss said when we were hired, various comments about the last person who held the job, the supervisor's reactions to the last project we completed, and so on. We concentrate on the relevant details in whatever order they make sense. Thus, if we narrate an incident in chronological order, we may leave it up to our audience to reorder the details in a meaningful fashion. We do not make or express our judgment about which details are the most important or what conclusions can be drawn. If we narrate an incident in developmental order, we have decided upon the meaning of the incident and narrate the details in the best order to illustrate that meaning.

Narrative then presents a subject in motion. This motion or action is framed by time. The time frame and sequence in which we narrate give the action structure and meaning. There is a point to our narrative and the order in which we tell it can support that point. Let us look further at time in narrative.

## Time

Although we can measure how long each stage of a real incident took in minutes, hours, days, and so on, we may present many or few details about each of these stages. The more detail we give to a stage, the more attention the audience pays to that stage and the more important it seems. For example, let us say that in an incident, stage A took 20 minutes to happen, stage B 24 hours, stage C 3 hours, and stage D one year. We decide that stage C is the climax of the incident, the event that illustrates the overall meaning; the actual stage C took a short amount of time but we want to tell a lot about it. Stage A might be another meaningful incident, and stages B and D less important in recreating the incident. The sequence, time frame, and selection of detail of our sample narrative might look like Table 8-1.

**TABLE 8-1.  Sequence, Time Frame, and Detail of Sample Narrative**

| Actual Incident | | Recreated Incident (Narrative) | |
|---|---|---|---|
| Stages | Time | Stages | Detail and Importance in Narrative |
| A | 20 min | B | very little |
| B | 24 hr | A | much |
| C | 3 hr | C | very much |
| D | 1 yr | D | little |

Our narrative might be something like the following:

John Smith was born in Brooklyn, N.Y., in 1965. His mother was in difficult labor for 24 hours, a full day and night.   *Stage B*

Shortly before John's birth, his mother fell down a short flight of stairs. She sustained bruises on her arms and legs. She was alone at the time and took a while to pull herself up and call for help. She didn't believe that she had injured her unborn child, but shortly after she went into labor. She was 8½ months pregnant. The baby appeared to be normal and healthy.   *Stage A*

When John was in the second grade, he was playing on a swing set during recess. He and a friend were swinging sideways, bumping into each other with their swings. John dared his friend to play the game without holding onto the swing chains. The boys raised their arms above their heads and swung at each other violently. John fell off his seat and hit the soft sand beneath. He hit the ground first with his side and then with his head. He stood up a moment, muttered a few words, and then sank to the ground unconscious. When he regained consciousness 3 hours later, he appeared to be fine. Tests revealed no injuries.   *Stage C*

During the next year, John began having headaches and problems with vision. Although it seems logical that John's accident and his mother's fall might have started his problems, tests and examinations could not confirm this hypothesis.   *Stage D*

In our sample narrative, we sequenced the stages of the incident to show the relationship between them while making the narrative interesting and meaningful for the audience. Also, although stages A and C took less time to happen in real life, we present more detail about them because of their importance. Stage D may draw a conclusion, but the real essence of the incident is recreated in stages A and C. A narrative then is a recreation of an event in a time frame that illustrates the meaning or essence of that event.

## Selecting a Beginning and an Ending of a Narrative

We can also see that in all narrative, we have a beginning, a middle, and an end to our recreated event. The beginning may describe a situation and include very little action. The beginning may be *exposition*—preliminary information necessary to understand what happened: the character's name is John Smith, and he was born in 1965 after a difficult and long labor (stage B). The beginning should have enough information to capture the audience's interest and perhaps hint at the action to come.

The middle of the narrative is the main body of the action,

perhaps the climax of the narrative. John Smith's mother fell at the end of her pregnancy (stage A) and after John fell during childhood, he remained unconscious for a long time (stage C). The situation or information given in the beginning becomes important and more complicated in the body of the narrative. Most of the action is given in the body where the audience "sees" more movement and detail.

The end of the narrative closes the action. All the complications have worked out or the final result of the action becomes clear. Through a series of accidents, our character John Smith may have suffered brain damage which could not be detected but which affected his health (stage D). The narrative ends at the point where the audience senses the full meaning of the action. The narrative then is complete.

In much technical communication, we have two choices of overall organization similar to those in narrative sequencing. We can state conclusions and recommendations first or last. In other words, we can lead an audience to a conclusion step by step and hope that it agrees with our conclusion at the end, or we can state our findings and then prove or illustrate them. Leading up to a conclusion is much like writing a mystery story in which the villain is revealed in the end; the audience will probably read to the end of the story but may not notice all our clues. On the other hand, using the same analogy, if the villain is revealed in the beginning and then the crime narrated, the audience sees the clues along the way but the narrative has to be important enough for the audience to want to finish it.

# Narrative to Introduce a Technical Communication

We use narrative in many types of technical communication: accounts of experiments, histories of scientific and technological progress, autobiographies and biographies, case histories, and accounts of such incidents as the Three-Mile Island accident. We often use narratives of experiments to introduce reports of scientific and technological progress. We also use narrative to introduce the past history of an issue, the literature previously published on a topic, or company growth or progress in a certain direction. These narratives may give important background on a subject so that an audience can judge how new projects may affect research or company progress. For example, the following is an introduction to a communication recommending an insurance plan to pay for second opinions on elective surgery:

> There is growing interest on the part of managers of union welfare funds and insurance plans in a program to cover the cost of obtaining a second opinion when nonemergency surgery is recommended by the patient's doctor. Such a program could result in a reduction of unnecessary surgery as well as related medical and hospital expenses.

One of the first second-opinion programs was adopted by Union A approximately 10 years ago. Under that plan, it is mandatory that a second opinion be obtained before consideration be given to paying the surgical or hospital expenses. Of course, this requirement does not apply to emergency surgery. Officials feel that a mandatory program gives the member an excuse to use when he or she confronts the doctor with the fact that he or she is seeking confirming opinion. In this case, the plan administrators have a list of board-certified specialists, a list that they have developed over the years. From that list they select the doctor who will be asked to give the second opinion.

In March 1972, Union B established a second-opinion program. Their program is also mandatory.

In August 1972, Union C established a second-opinion program. Their program is entirely voluntary. They use the approach that the program provides members with the highest quality of medical care in determining the need for surgery.

Both Union B's and Union C's plans were established with the help of a public health specialist who made arrangements to have most of the confirming diagnoses made by area hospitals.

Early in 1975, Union D established a plan in Maine. Their plan, as well as all other plans currently being established, are on a voluntary basis. This plan was designed jointly by the Director of Health Services of the Union and the General Director of a major hospital.

A clear introductory narrative helps the audience visualize completely the history of the subject. The narrative brings the audience up to date on developments and establishes a common background between the writer and the audience.

# Narrative to Organize a Technical Communication

We can also use narrative to organize an entire technical communication. In progress reports, trip or investigation reports, and periodic reports, we narrate what has happened over a set period of time. In these communications we usually report upward to a superior. For example, in the following progress report, the writer narrates a visit to a supplier and tells how his findings affect an ongoing project:

This is a progress report since our last memorandum of October 13, 1978. We have ceased exploring the first four options and are now concentrating on the fifth option, which is the installation of A.B.C. motor operators on North and South doors.

On November 9th, John Long, Fred O'Hanlan, Dennis Fisher, and I visited the A.B.C. factory in New York. We witnessed laboratory demonstrations and explored the feasibility of installing the A.B.C. motor operators for the southeast pair of doors as a trial in Percy Hall. You will recall that we have been in conversation with A.B.C. over this possibility since September of last year. Many of the bugs that showed up in their research and development effort had been worked out prior to our trip to the laboratory.

In our viewing of the operation, it was operated five times without any hitch. The sixth time the chain came off and it was deemed advisable to add a guide that would bring the chain onto the sprocket in the same plane in which the sprocket was operating. A simple guide can be developed to resolve this problem. We put in a temporary guide to so indicate and all in attendance agreed to the solution.

Those present indicated to A.B.C. that we are ready for the installation. They will ship all of the new rail and equipment to Kansas City the middle of December for installation between January 16 and January 20, 1979.

There is a change in scope in the contract from that which we understood last year. That involves additional materials in the form of a new traveling rail which has a curved slot in it, compared to the right-angle slot. Their installation will remove the existing rail and install the new rail. The operation is very simple, with a ½ hp motor driving a sprocket, which operates a chain, which pulls the door out of the closed position into a storage position, to give the full 30-foot opening. When the installation is complete all will be advised.

The writer recreates for his supervisor what happened when he visited a potential supplier. By reading what happened, the audience can almost "witness" what the writer has seen. The writer "reorders" time to emphasize progress made.

Whether we use narrative to *introduce* a communication or to organize an entire communication, we recreate for our audience an event or series of events. We choose a point of view, details, and a sequence which give the essence of the events.

# Causality in Narrative

Although when we narrate an incident, we are concerned primarily with "what" happened, to give the audience a full picture we might include the "why" or cause of the incident. However, in narrative, the cause is secondary to recreating the event. As we saw in Chapter 6, when causal analysis is our basic concern, we *analyze* rather than *relate* what happened. In narrative, causality might become clear as the story unfolds, but our main concern is still with the "unfolding." The "why" of the incident may give depth to the "what." Why the Three-Mile Island reactor failed gives a clear picture of the failure itself, but in narrative we discover the reasons for such a failure as we "watch" the reactor being shut down.

# Style in Narrative

Since narrative recreates a scene for an audience, the words we choose must help the audience *sense* what happened. Our words can involve the audience by telling how something sounded, tasted, felt, and so on, while the incident took place. However, although adjectives describe sense impressions, in narrative we are most concerned with action words—verbs. Our verbs must be vivid and descriptive to help create for an audience a sense of being there: "the reactor scrammed"; "its pressure rose"; the queen "withdraws her sting." Our verbs in narrative must be precise and vivid.

Since in narrative we recreate something that has already happened and may not happen again, we use past tense of verbs. In some special types of narrative that we study next, we use different verb tenses.

Finally, in narrative we must indicate time. When the incident began and ended, when changes occurred, and how long the stages of the incident took are important. We must frequently remind the audience how much time has passed.

## Summary

In narrative we "recreate" an incident for an audience. We present action, events in time and motion. Narrative recreates an event by giving the audience a sense of what happened. We select a point of view for our narrative. Our choice influences whether our audience is concerned with participants' or observers' thoughts as well as actions. If we relate our narrative from the first-person-participant point of view, our audience finds our thoughts and actions interesting; if we relate our narrative from first-person-observer point of view, our impressions of others' actions are important. We can relate a narrative from the third-person point of view by revealing the thoughts of all "characters," by revealing only the thoughts of the main character, or by relating objectively only observable actions.

We select detail to include in narrative according to our audience and purpose. We can also order our narrative chronologically and leave our "conclusions" for the end, or we can develop it to illustrate a certain point. How much detail we relate at each stage in the incident may indicate how important that stage is, rather than how long it actually took. At what stages we begin and end also reflect what is important in the incident.

If we indicate cause in narrative, it is secondary to what happened. Narrative gives an audience a sense of action through verbs within a time frame. In narrative when and what happened are essential.

In technical communication, we use narrative to present accounts of experiments and incidents, histories or scientific and technological progress, autobiographies and biographies, and case histories. We often introduce the background of our subject through narrative. We can also use narrative to organize a communication; we can place an investigation or a trip within a time frame to help our audience "see" what we saw.

# Readings
## to Analyze and Discuss

## I

The infant girl was the first child of a gravida 1, para 1, 20-year-old single Caucasian mother of upper-middle-class background who was a university student. The mother had received regular prenatal care since the third month of pregnancy. The option of terminating the pregnancy was offered at the time of the first medical consultation but was refused.

Pregnancy was uncomplicated. The date of the last menses was February 11, and delivery occurred on November 12. The infant was full-term at birth and weighed 2,898 gm. Delivery was spontaneous with vertex presentation and meconium (2 +) was noted. The infant was 50 cm in length and had an occipital frontal circumference (OFC) of 35.5 cm. Her blood type was 0 +. Apgar score was 9 at one and five minutes. The findings of examination at birth were normal. The infant was heard grunting two hours after birth. Breast-feeding was successfully begun.

The mother and infant were discharged from the hospital 48 hours after delivery (customary at the Medical Center). The next weeks were uncomplicated. The child was asymptomatic and doing well until the mother noticed an inspiratory "noise" (stridor) during sleep. She also mentioned, retrospectively, that the child had become "blue" at irregular intervals during sleep between 4 and 5 weeks of age. After the fourth week of life, the infant was reported to be gaining weight slowly although no pathologic condition was apparent. The mother noted that the sleep-related inspiratory "noise" seemed to worsen with age. At 2 months of age breast-feeding was discontinued and feeding with a commercial formula was begun and was followed by increased weight gain. At 12 weeks of age, the infant had a "runny nose" and congestion. The mother noted that the baby breathed more rapidly, stridor had increased, and she seemed "blue" on two occasions. The infant was hospitalized 48 hours after the onset of clinical symptoms. She was small for her age, weighing 3.9 kg. She had loud inspiratory stridor and mild respiratory distress. Rectal temperature was 37.7 C. . . .

The infant improved during one week of observation in the hospital, and mild inspiratory stridor was noted only during sleep. She was discharged with the diagnosis of "congenital stridor, etiology unclear." During the following weeks, the mother again noticed episodes of cyanosis during sleep. At 4½ months of age she was referred to the SIDS research program for polygraphic monitoring and was rehospitalized for the study. On physical examination, she weighed 5.96 kg, was 56 cm in length, and had an OFC of 44.2 cm. Inspiratory stridor was observed during crying and sleeping. No nasal obstruction was apparent, and she could feed without difficulty. The infant was discharged from the hospital, after completion of 24-hour monitoring, on a Friday afternoon. She was last observed alive in her crib by the mother at 4 A.M. Sunday and was found dead in her crib about 8 A.M. Sunday morning.

## Questions for Discussion

1. What are the audience and purpose of the narrative above?
2. What is the point of view? Why did the writer choose it?

How much detail does the writer relate about each stage of the narrative? Why?

3. Does the writer reorder time? Is the sequence of the narrative chronological or developmental? Why?

4. Why did the writer choose to begin and end the narrative as represented here? Is causality given in the narrative?

# II

I was experimenting at one time with young mallards to find out why artificially incubated and freshly hatched ducklings of this species, in contrast to similarly treated greylag goslings, are unapproachable and shy. Greylag goslings unquestioningly accept the first living being whom they meet as their mother, and run confidently after him. Mallards, on the contrary, always refused to do this. If I took from the incubator freshly hatched mallards, they invariably ran away from me and pressed themselves in the nearest dark corner. Why? I remembered that I had once let a muscovy duck hatch a clutch of mallard eggs and that the tiny mallards had also failed to accept this foster-mother. As soon as they were dry, they had simply run away from her and I had trouble enough to catch these crying, erring children. On the other hand, I once let a fat white farmyard duck hatch out mallards and the little wild things ran just as happily after her as if she had been their real mother. The secret must have lain in her call-note, for, in external appearance, the domestic duck was quite as different from a mallard as was the muscovy; but what she had in common with the mallard (which, of course, is the wild progenitor of our farmyard duck) were her vocal expressions. Though, in the process of domestication, the duck has altered considerably in colour pattern and body form, its voice has remained practically the same. The inference was clear: I must quack like a mother mallard in order to make the little ducks run after me. No sooner said than done. When, one Whit-Saturday, a brood of purebred young mallards was due to hatch, I put the eggs in the incubator, took the babies, as soon as they were dry, under my personal care, and quacked for them the mother's call-note in my best Mallardese. For hours on end I kept it up, for half the day. The quacking was successful. The little ducks lifted their gaze confidently towards me, obviously had no fear of me this time, and as, still quacking, I drew slowly away from them, they also set themselves obediently in motion and scuttled after me in a tightly huddled group, just as ducklings follow their mother. My theory was indisputably proved. The freshly hatched ducklings have an inborn reaction to the call-note, but not to the optical picture of the mother. Anything that emits the right quack note will be considered as mother, whether it is a fat white Pekin duck or a still fatter man. However, the substituted object must not exceed a certain height. At the beginning of these experiments, I had sat myself down in the grass amongst the ducklings and, in order to make them follow me, had dragged myself, sitting, away from them. As soon, however, as I stood up and tried, in a standing posture, to lead them on, they gave up, peered searchingly on all sides, but not upwards towards me and it was not long before they began that penetrating piping of abandoned ducklings that we are accustomed simply to call "crying." They were unable to adapt themselves to the fact that their foster-mother had become so tall. So I was forced to move along, squatting low, if I wished them to follow me. This was not very comfortable; still less comfortable was the fact that the mallard mother quacks unintermittently. If I ceased for even the space of half a minute from my melodious "Quahg, gegegegeg, Quahg, gegegegeg," the necks of the ducklings became longer and longer, corresponding exactly to "long faces" in human children—and did I then not immediately recommence quacking, the shrill weeping began anew. As soon as I was silent, they seemed to think that I had died, or perhaps that I loved them no more: cause enough for crying! The ducklings, in contrast to the greylag goslings, were most demanding and tiring charges, for, imagine a two-hour walk with such children, all the time squatting low and quacking without interruption! In the interests of science I submitted myself literally for hours on end to this ordeal. So it came about, on a certain Whit-Sunday, that, in company with my ducklings, I was wandering about, squatting and quacking, in a May-green meadow at the upper part of our garden. I was congratulating myself on the obedience and exactitude with which my ducklings came waddling after me, when I suddenly looked up and saw the garden fence framed by a row of dead-white faces: a group of tourists was standing at the fence and staring horrified in my direction. Forgivable! For all they could see was a big man with a beard dragging himself, crouching, round the meadow, in figures of eight, glancing constantly over his shoulder and quacking—but the ducklings, the all-revealing and all-explaining ducklings were hidden in the tall spring grass from the view of the astonished crowd.

## Questions for Discussion

1. Why did the writer choose to narrate in first-person-participant point of view? How is this point of view appropriate to his audience and purpose? What is the writer's viewpoint—how does he perceive himself and the event he narrates?

2. Is the narrative ordered developmentally or chronologically? How can you tell? Trace the sequence of events as they actually happened and as the writer narrates them.

3. What are the audience and purpose of the narrative? How would the narrative have to be changed if presented to a group of eminent scientists at a professional conference? A group of college students? An environmental protection agency? A group of hunters?

# Exercises and Assignments

1. **Problem:** Our choice of point of view in narrative often influences our audience's impressions of the incident depicted. Reread "The Satellite That Invaded a Campsite" in the readings section of Chapter 1. Determine the point of view used and decide why the writer chose that point of view and what effect it has on the audience. Then do the writing assignment that follows.

Assignment:

Recreate for an audience of your choice the incident depicted in "The Satellite That Invaded a Campsite" from the viewpoint of the campers. Use either the first-person-participant, third-person-limited, or third-person-omniscient point of view. You may add detail to the incident if you wish.

2. **Problem:** Reread Reading I at the end of this chapter. The impressions of the mother would surely be different from those of the narrator of this anecdote.

Assignment:

Retell the narrative from the viewpoint of the mother.

3. **Problem:** As we have seen, we can narrate the stages of an incident or event chronologically as they actually took place or developmentally to illustrate a point or emphasize one stage in the event. If we narrate chronologically, we lead our audience to a conclusion, while if we narrate developmentally, we may indicate our conclusion early or we may interrupt our narrative with flashbacks or anecdotes.

Assignment:

Recreate an important event in your life or an important scientific or technological discovery through narrative. For example, you might tell of your first week in college, or you might relate Ben Franklin's first experiments with electricity (you would have to do research here). (1) Narrate the event first chronologically, as it actually happened, and then (2) narrate the event developmentally by first relating the most significant stage in the event and then giving the background leading up to that stage and the results of it. Designate appropriate audiences and purposes for both narratives.

4. **Problem:** You are an elementary school science teacher who wants to explain the events surrounding the Three-Mile Island nuclear incident to sixth-grade students. Your sixth graders have heard of the incident on television news and have an elementary knowledge of nuclear fission.

Assignment:

Reread the two narratives that relate to incidents at Three-Mile-Island and write a narrative of these incidents that your sixth-grade students could understand. Choose an appropriate point of view and amount of detail as well as an interesting sequence.

# Process Description:
## Describing Devices or Systems in Operation

*Including the human agent in process description.*
*Explaining the principles of operation in a process.*
*Using process description to describe an operation to a technical and a general audience, to explain a new technical system, to speculate about scientific or technical theory, and to present alternatives.*
*Supporting process description with graphic aids.*
*Writing effective sentences in process description.*
*Reading and analyzing process descriptions written by others.*

In the last chapter, we studied how to present a subject in motion. Freezing a subject in time and motion is a good way to get a close look at it, to define it, to describe it, to classify it, or to compare it with or contrast it to another. However, just as after we meet someone for the first time, we want to know what that person does for a living, how he feels about an issue, or how she reacts to other people, after we see the parts of a device, we want to see that device at work. So we use a special type of narrative to describe systems or devices in operation—process description.

After we have described what a device looks like, usually we describe what it looks like in motion, by what principles it operates, how one part affects another, and the final effects of all these parts or causes in motion. As we saw with narrative, to tell our audience how our subject moves, we sequence movements in time: first this happens and then that. In process description, we combine narrative and causal analysis: first this happens which then *causes* that to happen. Of course, the effects in this process are deliberate; the parts were designed to move other parts: "this" was designed to cause "that" to happen. A human being may be just one of the "parts" in a process, and the process may operate according to a certain scientific principle or law.

For example, in 1897 Sir J. J. Thomson conducted experiments to discover the nature of cathode rays. In a report published in *Philosophical Magazine,* Thomson first described the device he used in the experiment and the basic operation of this "electrometer." In the following passage, Thomson describes the electrical process involved:

> When the rays are turned by the magnet so as to pass through the slit into the inner cylinder, the deflexion of the electrometer connected with this cylinder increases up to a certain value, and then remains stationary although the rays continue to pour into the cylinder. This is due to the fact that the gas in the bulb becomes a conductor of electricity when the cathode rays pass through it, and thus, though the inner cylinder is perfectly insulated when the rays are not passing, yet as soon as the rays pass through the bulb the air between the inner cylinder and the outer one becomes a conductor, and the electricity escapes from the inner cylinder to the earth. Thus, the charge within the inner cylinder does not go on continually increasing; the cylinder settles down into a state of equilibrium in which the rate at which it gains negative electricity from the rays is equal to the rate at which it loses it by conduction through the air. If the inner cylinder has initially a positive charge it rapidly loses that charge and acquires a negative one; while if the initial charge is a negative one, the cylinder will leak if the initial negative potential is numerically greater than the equilibrium value.

Thomson tells his audience what happens within a time frame: "when the rays are turned" and when they "pass through" the bulb. He uses precise verbs to describe action: "turned," "escapes," "settles down," and "loses that charge and acquires a negative one." The phrase "this is due to the fact" announces that the sequence is also causal.

We see that process description indicates motion and time, as well as cause. In this chapter, we study how to select details and sequence movement in process description as well as how to indicate cause.

All the choices we make when writing a narrative pertain to process description. We choose a point of view, the details to include, and the sequence in which to present these details. However, since in process description we describe a device or system in operation, we must keep in mind not only our audience but also our special purpose when making these decisions. We review these decisions as we study the special characteristics of process description.

# Selection, Sequence, and Causation in Process Description

Usually we sequence process description chronologically to help our audience understand the normal order of the process. Unlike narrative, which depicts an event that may have occurred only once, process description relates how a device works every time it is set in motion. We seldom reorder the natural sequence of that motion. Chronological order is also necessary to depict the causes of the motion. Although several parts may contribute to one action, usually each effect becomes the next cause; each motion creates another. When we describe a process then, we describe each motion in turn and the results of that motion. We select detail according to our audience and purpose.

## An Example of Process Description

In the following process description, the writer describes the operation of a toilet. She first gives a preview of the process description and the principles involved. She then describes each step in the process and the effect of each step:

> The most often used plumbing device in the home, and therefore the most susceptible to breakdown, is the toilet. The toilet is engineered to remove human waste in a very precise manner. The principle of operation of the toilet depends on the natural phenomena that submerged and detached air bubbles rise. The lever principle is also responsible for the successful working of the many internal parts of the toilet. The ordinary toilet performs its flushing in two major processes: first, the flushing of the bowl which removes the waste, and second, the refilling of the tank and toilet bowl.
>
> The first step of the flushing action is initiated by the person pushing the flush handle in a swift counterclockwise circular direction. The operation continues spontaneously after this initial input of energy, because each piece of equipment is integrally connected to the other working components. Moving the flush handle on the outside of the tank actuates a metal piece directly connected to the handle on the inside of the tank. As the metal piece is abruptly lifted, it elevates

the stopper wire connected at right angles to it. These stopper wires then pull the stopper ball up. As the stopper ball is pulled away from its seat, water pent up in the tank rushes through the discharge pipe into the toilet bowl which causes the wastes in the bowl to flow through the waste pipes into the sewage system. The stopper ball rushes upward rapidly through the water because, once ajar from its seat, the stopper is very similar to a loose air bubble.

The second process, which regulates the water level of the tank, involves a number of steps. The first step is the steady descent of the rubber float, a rubber hollow ball, as the water level subsides slowly in the tank. The rubber float always floats on the surface of the water in the same way that a closed bottle floats in water. As the rubber float goes down with the water level, the float arm attached directly to it moves and opens the water inlet valve. Water streams slowly into the tank at a much slower rate than water leaves through the discharge pipe. Simultaneously, from the same water inlet valve a small spray of water is squirted through the refill tube into the outflow pipe and finally, into the toilet bowl where a full bowl of water is maintained. The spray of water also serves as a precautionary device. If the water inlet continues to flow into the tank although it is full, the extra water will simply flow into the refill tube through the outflow pipe into the toilet bowl and out through the waste pipes. Inside the tank the water level falls and the stopper ball floats on the lowering water surface until it reaches its seat. It stays stuck to the seat as rising water exerts a downward pressure. Since there is no longer a flow of water out of the tank, the float moves upward with the rising water level in the tank. Once the float has risen to its equilibrium position, the water inlet valve is shut. No water is forced into the tank or refill tube. The filled tank is now ready for another flush.

*Kim Dellas*

Notice that the writer reminds her audience constantly of the time frame of the action: "as soon as" one step is complete, another begins; "simultaneously" two actions occur; "once" an action ends, another begins. The writer balances action, time, sequence, and cause. The audience knows what step occurs, when, before and after what other step, and what causes it to occur. The writer depicts a process that involves continuous action for an audience that needs to know how the device operates. She describes each change or step in that operation so that her audience can visualize a complicated process.

In process description then, we sequence actions in time. Because the same process usually occurs every time the device is set in motion, or turned on, we order the process description chronologically. This order enables us to describe the steps in the process as causes and effects. Since we describe action rather than thoughts and feelings, we usually use the third-person-objective point of view in process description.

## The Human Agent in Process Description

If a human being plays an essential role in the process we describe, we include this "operator" or "agent" as one of the parts of the process. Since in a process description we describe the action so that our audience can understand it rather than necessarily repeat or participate in it, the members of our audience are often not the operators or agents described in the process. In process description, the agent or operator is just as but not more important than one of the other "parts" of the device in motion.

## Principles or Laws in Process Description

To help an audience understand how a device works, we also explain what scientific principles or laws, if any, are at work during the process. For a technical audience we can simply name the law; for a nontechnical audience, as addressed in a communication directed outward or upward, we can explain the principle; for example, if the process involves a lever, a small force that acts over a large distance is converted to a large force that acts over a small distance. An audience can more easily understand a process if we explain what scientific law makes the process possible.

For example, in the following process description which appeared in a technical bulletin, the writer explains the principle involved in the operation of temperature controllers—Charles's Law:

> All the Wizard II® temperature controllers (Fig. 9-1) accept, as an input, the process temperature that is sensed by a temperature bulb immersed in the process fluid. The temperature bulb, a capillary tube, a Bourdon tube, and a temperature gauge calibrated for the appro-

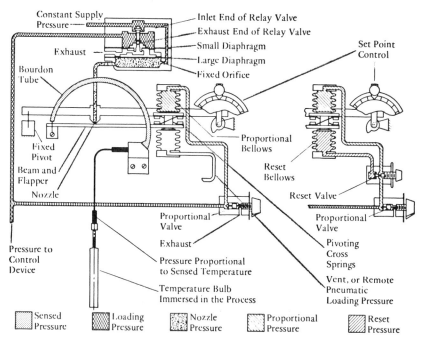

*Figure 9-1.* Courtesy of Fischer Controls.

priate temperature range, form a closed system referred to as the thermal element assembly. The capillary tube connects the temperature bulb to the Bourdon tube and the temperature gauge (both are inside the controller case).

The operation of gas-filled temperature bulbs is based on Charles's Law, which says that if the volume of a given weight of gas is kept constant, the pressure of the gas will vary directly with the absolute temperature. Temperature bulbs used with the Wizard II® controllers contain gas-filled charcoal that emits gas as the process temperature increases and absorbs gas as the process cools. Thus, an increased process temperature builds pressure within the Bourdon tube, and decreasing pressure diminishes it.

Because the volume of the temperature bulb is much larger than the volume of the capillary tube, temperature errors caused by the ambient temperature of the capillary tube are negligible.

After this overview of the principle behind the process, the writer described the device in operation. Stating the principles behind the process lets our audience draw a comparison with a similar yet universal process.

# Audiences and Purposes of Process Description

Often we present a process description to an audience after we have described the device and before we give instructions on how to use or fix the device. Process description can also help an audience buy a device, understand a new scientific or technical theory, choose between alternative actions, or understand company action or policy.

## Process Descriptions for the Potential Customer

Technical manuals, catalogs, and instruction sheets accompany most devices we use on the job. We read these communications before using or buying the device, and we write them if we are selling, manufacturing, or developing the device. A detailed process description usually addresses a technical or semitechnical audience. The following process description preceded an explanation of the device's capabilities. The process description convinces the potential user that the valves can reduce cavitation noise:

> As a result of Fisher research, flow conditions that will produce damaging cavitation can be accurately predicted. Cavitation and its associated noise and damage can often be avoided at the design phase of a project if proper consideration is given to service conditions. However, when service conditions are fixed, a valve may have to operate at pressure conditions normally resulting in cavitation. In such instances, noise control by source treatment can be employed by utilizing one of several methods: multiple valves in series, a special control valve, or a standard control valve body with special internal parts.

Cavitrol® III and Cavitrol IV trims reduce cavitation noise and damage by taking the total pressure drop in a series of intermediate stages. In operation, fluid enters the first section of the cage through many orifices. In passing through orifices, each fluid stream undergoes a portion of the total pressure drop. The fluid then passes through a series of additional orifices and undergoes additional pressure drops at each stage. The number of stages required to prevent cavitation depends upon the total amount of pressure reduction that must be taken across the cage. A control valve using Cavitrol III or IV trim will exhibit a noise level of 90 dBA or less.

The audience of this description would have enough knowledge and authority to buy the device; however, because the writers describe action clearly within a time frame, any audience can understand the basic process involved.

## Process Descriptions for the Public

Process descriptions also address nontechnical audiences. For example, the following process description appeared in a public relations release. Upon the opening of a new plant, the company released information to area newspapers so that visitors or interested local people could learn about and appreciate the plant operation. In one of a series of releases, writers described the operation of water systems and air and water environmental controls of the new "L" blast furnace. Notice that the writers combine process and device description. Although the process description is thorough, the writers use common words and images: for example, dust tubes work like "vacuum cleaners." Even though terms like "baghouse" challenge someone who has never visited a steel plant, the process of operation is clear. The description helps the audience understand a process they might never see or be involved in:

Fumes generated during the casting operation are collected from the casthouse floor through a suction main under the floor that draws fumes from above the iron and slag tilting spouts which are topped by removable covers during operations. Overhead draft hoods at the tap holes increase the effectiveness of the system.

Evacuated by a 312,000-cubic-foot-per-minute fan driven by a 1,500-horsepower motor, the fumes are passed through a five-cell baghouse that contains 3,360 Dacron dust tubes, each one 22-feet long and 8 inches in diameter. The dust tubes, which operate on the principle of a home vacuum cleaner, are periodically shaken clean by electric-powered mechanisms.

Screw conveyors transfer the dust to trucks for disposal. Eventually, the dust, if found suitable, will be recycled in the steelmaking and ironmaking processes as is the case with almost all other Sparrows Point iron-bearing wastes.

Blast furnace gas is processed through a series of operations designed to clean 360,000 standard cubic feet per minute to a level of 0.005 grains per cubic foot. The gas first passes through a 41-foot, 6-inch dust catcher to eliminate heavier particles. It then flows through an installation of high-energy scrubbers and moisture separator that is unique in this country.

The scrubbers also serve the function normally handled by septum valves for decompression of the blast furnace gas and thereby regulate the furnace top pressure. This function is regulated from the control room.

The water from the gas scrubber system goes to two 90-foot-diameter parallel thickeners which settle out the solids for recycling. The cleaned water, after cooling in a two-stage cooling tower, is chemically treated and recirculated. The slurry recovered from the thickeners is pumped to the sinter plant for recycling. . . .

With new concerns about ecology and public safety, more than ever before, scientists and technologists must explain technical processes to the public. The explanations must be clear and accurate. Notice that the writers of this process description stress the recycling capacities of the new plant.

## Process Descriptions for Peers

We also describe technical processes that replace or improve old ones. Usually addressed to a peer or lateral audience, such a process description might appear in an in-house communication or a technical journal. For example, the following process description explains a new computer system. Notice that although the vocabulary is technical, the writer still indicates clearly action, sequence, and cause:

Commands to a task are received from the C/D FIFO buffer. A FIFO interrupt, recognized by the frontend PIO, triggers the executive routine which fetches the FIFO contents and transfers control to the task's command handler. The command handler then decides what type of action is required by examining the 16-bit data pattern of the command, making use of executive service if required, and takes that action. Control is then returned to the executive. Note that online task routines must execute as fast as possible in order to make way for other tasks which may be time dependent. In one design case, compliance with this general rule required that an interrupt routine be divided into two portions; the second half of this routine is triggered by programming a PIO to generate an interrupt when an unused output bit is written to the true state. This splitting of a low priority interrupt routine permits higher priority activity to intervene while guaranteeing that the second half will execute much sooner than if it were a polling routine. . . .

The audience addressed in this process description knows thoroughly how a computer works. The writer's task is to explain a new type of operation.

## Process Description to Predict or Recommend

Speculating as to what may happen in the future is another task of the scientist and technologist. We can use process description to predict possible occurrences for both technical and lay audiences. Here cause would be essential within the description. For example, in the following process description, the writer speculates about what happens to the stars as the universe expands. The principle or laws of gravity support this speculation. Here the devices or systems of operation described are the stars within the universe:

> Due to gravity the matter in any aggregation tends to collapse inward toward its center of mass. In a star this inward force is balanced by the heat and radiation released in its core by nuclear burning of hydrogen into helium and helium into heavier elements. When, after several billion years, the star has exhausted its nuclear fuel, its gravity will no longer be balanced and the star will start to contract. If its mass is less than about 1.4 solar masses, the star may settle down finally to become a white dwarf, composed of iron nuclei (iron has the most stable nucleus) from which the intense gravitational pressures have stripped all the electrons. These electrons move freely throughout the star and together exert a pressure that resists further contraction. A white dwarf of one solar mass is about the size of the earth. As they radiate away the remainer of their energy, white dwarfs very gradually cool and grow dimmer, becoming in the end "black" dwarfs.

Using scientific principle and observation, the writer goes on to describe how a black hole is formed. Causation and action in sequence help the audience imagine a process that cannot be seen.

Finally, we can use process description to support an argument. Here tight sequencing and transitions between actions are as essential in our description as logical thinking and transitions between ideas are in the argument. If our audience follows closely our process, they will agree with our analysis on how to improve or change that process. To accept a proposed change in a process, our audience must understand the process completely.

We use process description to help an audience understand a specific operation, decide whether to buy or use a device, speculate about future occurrences, or evaluate recommendations. Although the type and amount of detail we use in process description vary with the education and experience of our audiences, we must indicate clearly action, sequence, and cause.

## Graphic Aids in Process Description

Graphic aids such as flow charts and diagrams support process description. Diagrams of the machine or device show the movement or action between parts by such means as arrows. In fact, if a process description is simple and not the main purpose of the communication, we can indicate action within the diagram and describe process within a caption, as in Figure 9-2. Here the writer can refer to the parts or steps labeled within the diagram.

Usually a diagram can indicate only the main actions in a process. The arrows in the diagram help the audience follow the description. Although we must remember that graphic aids support rather than replace a description, clearly labeled flow charts or diagrams can represent visually the main steps in the process described. They can represent action and sequence although not time and cause (see Figs. 9-8 and 9-9, p. 179).

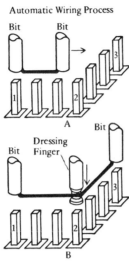

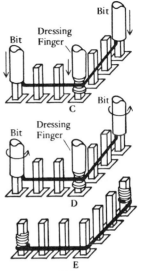

A "robot" wiring machine, manufactured for Western Electric by Gardner-Denver, automatically cuts and wraps all the wires for the solderless wrapped connections between terminals in the backplane of a 1A Processor.

To connect terminals one and three, for example, the Gardner-Denver machine positions a wire-wrap bit holding one end of the wire over terminal one (A). A second wire-wrap bit moves toward terminal three, unrolling the wire as it moves.

As the second bit turns the corner at terminal two (B), a dressing finger drops to hold the wire at the corner. The second bit then continues moving toward terminal three. When the second bit reaches this terminal, the wire is cut and stripped and the bits and dressing finger are lowered over the terminals (C). Then the two bits turn, twisting the wire around terminals one and three respectively (D). In the final step, the bits and dressing finger are raised, leaving the completed connection (E).

*Figure 9-2. Bell Laboratories Record*, Vol. 65, No. 11 (December 1978), 291. Copyright © 1980; Bell Telephone Laboratories, Inc. Reprinted with permission from the Bell Laboratories *Record*.

# Style in Process Description

As with narrative, verbs express the action in process description. Since a process *is* an action or movement, we must choose verbs that express the action exactly. We must check the intensity of the actions represented in our verbs. Do we mean "fall," "drop," or "decline"? Do we mean "separate," "disconnect," or "split"? Decline represents a slight descent, drop a great one, and fall a rapid one. Whole units separate from others in a group whereas parts of a unit may be disconnected, and random pieces split under pressure. We must be sure that we represent the exact action in our verbs.

Usually we use the present tense of verbs in process description. The present tense tells our audience that whenever the device is set into motion, the same things happen: plate mills roll plates, oxygen reacts with impurities, gasoline vaporizes. The present tense also gives our audience a feeling of immediacy.

To indicate sequence in process, we use adverbial transitions. We describe action as it occurs in time. Transitions such as "when," "as soon as," "immediately after," "at the same time as," and "then" not only connect the steps but also tell when they happen in relation with each other.

We must also explain clearly causation in process. We use phrases such as "because of," "due to," or "caused by" to indicate to our audience causal sequence. Unlike narrative, process description emphasizes the "why" as well as the "what." Starting with the scientific principle at work in the process helps our audience understand the overall "why" of the process.

Other stylistic guidelines depend on our audience and purpose. Technical audiences may need to know the exact degree of movement while general audiences may need only the basic action. A process description may follow a device description for a technical audience, while for a general audience we might indicate only the location of the parts while describing their action. If we describe a complicated process, we might preview the process by listing the steps in an introduction.

Finally, because we are describing a process as a *flow* of action, we might be tempted to use long, run-on sentences. Although shorter sentences may seem to break our description of that flow, we must give the audience time to pause and absorb our description.

# Summary

Process description, a special type of technical narrative, describes a system or a device in operation. Usually presented in the third-person-objective point of view and chronological order, it indicates motion in time and sequence and the causes of that motion. The detail we include in a process description depends on our audience and purpose. We include the human agent or operator in process description only as one step or part in motion. The purpose of process description is to help our audience understand but not necessarily duplicate an operation. Usually we indicate the scientific principle or law at work in a process to help our audience anticipate the action we describe.

In technical communications, we use process description to describe the operation of a device an audience might want to purchase and use. This description might appear in a technical manual, catalog, or instruction sheet. We describe technical processes to general audiences such as the concerned public when that process might affect the environment or life-styles. We describe new processes that result from research to technical audiences, and speculate on possible occurrences by emphasizing causal analysis in a process description. Finally, we can use process description to clarify and support recommendations.

We must also represent action accurately in our description by precise verbs in the present tense. Adverbial transitions indicate time, and phrases indicate cause. Graphic aids such as diagrams and flow charts best support process description. Process description then depicts for an audience a system or device in operation.

# Readings
# to Analyze and Discuss

# I

*Raw materials preparation* processes convert the iron ores and the coking coal into forms suitable for ironmaking.

The ores, essentially iron oxides containing some impurities, must be agglomerated into particles at least roughly the size of small marbles before they are introduced into blast furnaces. This is done either in pelletizing units (usually located at the mine site) or in sinterplants in which mixtures of fine ore particles, limestone, and other materials are made to undergo partial fusion.

The coke, which serves both as a reducing agent and as a fuel in ironmaking, is manufactured in coke ovens using metallurgical coal as a raw material. Finely ground coal is heated in the absence of air to temperatures at which volatile matter, moisture, and some of the sulfur compounds are driven off. The result is metallurgical coke; the volatiles, known as coke oven gas, are also a valuable fuel.

Most *ironmaking* is done in iron blast furnaces, which are in essence tall vertical shafts up to 30 meters high and 10 to 15 meters in diameter. Iron in ore, sinter, or pellets, flux (limestone), and metallurgical coke are charged at the top, while preheated air is blown into the system at the bottom. There is partial combustion of the coke, and the oxygen and minerals are removed from the iron ores, reducing them to metallic iron, during their descent through the furnace. The main products are molten iron (containing carbon, silicon, manganese, sulfur, and phosphorus), molten slag, and blast furnace offgas. Typically, a single such blast furnace produces some 1,000 to 10,000 tons of molten iron per day. . . .

The dominant steelmaking operation at present is the *basic oxygen* process, in which molten iron and scrap are introduced into a pear-shaped vessel and then a supersonic oxygen jet is blown onto the bath surface. The oxygen reacts with the impurities to create slag and offgas, and the refining is completed in less than 25 minutes, in contrast to several hours in the open hearth process. A variant of the basic oxygen process is the Q-BOP operation, which involves blowing oxygen and a coolant through the bottom of the vessel. . . .

One of these [changes proposed] is direct steelmaking . . ., in which iron ore is transformed to steel with no intermediate stage. This is proposed to be done by feeding a mixture of coal and iron ore into a molten steel bath, accompanied by oxygen injection. The reduction of iron oxide to iron would take place in the molten state, with the heat of reaction supplied by the combustion of the carbon with oxygen. Such direct processes are attractive because coke is not required, because cheaper grades of coal may be used, and because ironmaking and steelmaking steps are combined. Furthermore, the carbon monoxide that is produced is given up in a continuous manner so that its subsequent utilization can be more straightforward.

Direct steelmaking could save energy, capital and labor costs; no coke ovens or blast furnaces would be needed.

The principal problems to be overcome include the containment of the system, the difficulties asociated with the coexistence of oxidizing and reducing reactions within the same vessel, the feeding of the raw materials, and the management of start-up and shut-down. Some of these problems have resisted solution in a number of developmental tests, and it may be that ultimately no satisfactory resolution will be found; direct steelmaking is definitely not an established technology.

## *Questions for Discussion*

1. The purpose of this process description is to discuss and recommend technical solutions to problems in the steel industry. What was the audience? How can you tell?

2. Compare this process description to the one we examined earlier in the section on Process Descriptions for the Public. Can you explain the differences?

3. Basically, why does the writer recommend and not recommend direct steelmaking? Might the writer give this process a stronger recommendation in the future? Under what conditions?

4. Does the writer give his audience a sense of time, action, and cause in the process description? Which words or phrases convey these characteristics? What principles are involved in the process?

# II

The significant difference between nuclear and fossil plants is the fuel used by each. Most nuclear plants "burn" uranium-235 that has been mined, refined, and enriched (i.e. concentrated) and then molded into fuel pellets. When enough enriched uranium fuel pellets are brought together under the right conditions, uranium-235 atoms begin to split, or fission; this produces the heat needed to turn large quantities of water into steam.

In a fossil-fueled plant (see Fig. 9-3), coal, petroleum, or natural gas is burned. The heat of combustion is used to create superheated, pressur-

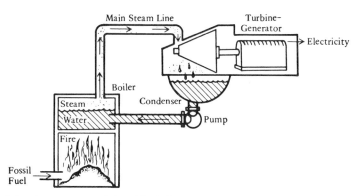

*Figure 9-3.* A conventional fossil-fueled electric power plant. Peter Faulkner, Ed., "How a Nuclear Reactor Works," *The Silent Bomb.* Copyright © 1977 by Friends of the Earth. Reprinted by permission of Random House, Inc.

ized steam in a boiler. The steam is piped to a tubine-generator, where it impinges on the blades of the turbine, causing it to turn. The turbine is connected to the generator by a shaft; the turbine-generator turns at high speed as high-pressure steam pushes against the turbine blades. When the generator turns, electricity is generated.

The spent steam passes from the turbine to a condenser through which cooling water is pumped from an outside source. The condenser absorbs some of the steam's heat, converting it again to water, which is pumped back into the boiler. Notice that the water coolant shown in Figure 9-3 circulates in a continuous "loop"—from boiler to turbine to condenser and back to the boiler.

A nuclear-fueled plant is diagrammed in Figure 9-4. When the nuclear fuel in the reactor core fissions, heat is produced, which boils water circulating through the reactor core. This design . . . is called a boiling-water reactor (BWR). Great quantities of steam at approximately 1,000 pounds per square inch (psig) pressure are piped to the turbine-generator, causing it to turn. The spent steam is condensed and returned to the reactor core as "feedwater."

Although the steam is radioactive as it leaves the core, most of the radioactivity is short-lived and does not heavily contaminate the turbine or feedwater systems. The BWR is the simplest of operating reactor system designs. It features a single steam-water loop through which steam is piped directly from the reactor to the turbine. The BWR also features control rods to control the rate of nuclear fission; these rods are inserted from *beneath* the core by hydraulic drives electrically activated by operators in the control room. . . .

The pressurized water reactor (PWR) diagrammed in Figure 9-5 is similar to the BWR design, but whereas the BWR circulates steam from the reactor to the turbine and back to the reactor through a direct "loop," the PWR features *two* loops. In the primary loop, pressurized water carries heat from the core to a steam generator, or boiler, where the heat is transferred through tube walls to water flowing through a second loop. Secondary-loop feedwater from the condenser is pumped into the steam generator, absorbs heat from the first loop, and turns to steam. The second loop carries the steam at about 1,000 pounds per square inch pressure to the turbine-generator and then to the condenser.

Note the similarity between the PWR and BWR. The PWR's secondary loop is designed to keep the turbine and associated pipes and pumps as free of radioactivity as possible. Thus the PWR's heat exchanger acts as a "buffer" between the two loops. If the tubes did not leak, the turbine, con-

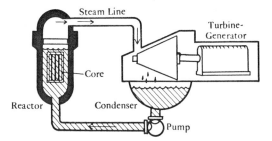

*Figure 9-4.*   Boiling water reactor (BWR). Peter Faulkner, Ed., "How a Nuclear Reactor Works," *The Silent Bomb.* Copyright © 1977 by Friends of the Earth. Reprinted by permission of Random House, Inc.

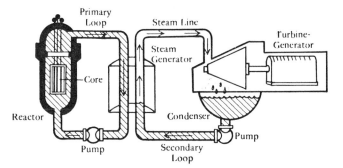

*Figure 9-5.* Pressurized water reactor (PWR). Peter Faulkner, Ed., "How a Nuclear Reactor Works," *The Silent Bomb.* Copyright © 1977 by Friends of the Earth. Reprinted by permission of Random House, Inc.

denser, pumps and piping would not become radioactive and thus would be easier to replace and repair. Unfortunately, nearly all steam generator tubes do leak, and contamination of the secondary loop is inevitable in these cases. Pressurized water reactors have control rods inserted from *above* the reactor core. . . .

The third reactor design examined here is shown in Figure 9-6. The liquid metal fast breeer reactor (LMFBR) uses molten sodium instead of water as the reactor core coolant. It is a more complex design than the BWR or PWR. The LMFBR is known as the "breeder" because it produces more plutonium fuel than it burns. . . .

Note the three loops in Figure 9-6. Molten, radioactive sodium circulates in the primary loop through a heat exchanger which transfers primary-loop heat to molten sodium (nonradioactive) in the secondary loop. Secondary-loop heat converts water in the *third* loop to steam as it circulates through a steam generator. The nonradioactive steam is piped through the third loop to drive the turbine generator. . . .

Figure 9-7 is a simplified diagram of the nuclear fission process. An atomic particle, a neutron in this case, strikes the nucleus of a uranium-235 atom, which fissions, or disintegrates, into fission fragments, heat, and one or more free neutrons. All this takes place in less than $\frac{1}{1000}$ of a

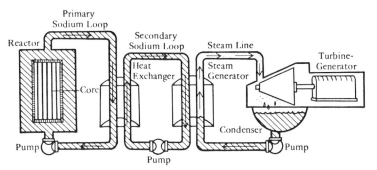

*Figure 9-6.* Liquid metal fast breeder reactor (LMFBR). Peter Faulkner, Ed., "How a Nuclear Reactor Works," *The Silent Bomb.* Copyright © 1977 by Friends of the Earth. Reprinted by permission of Random House, Inc.

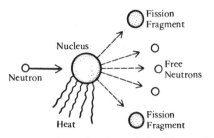

*Figure 9-7.* A fission reaction. Peter Faulkner, Ed., "How a Nuclear Reactor Works," *The Silent Bomb.* Copyright © 1977 by Friends of the Earth. Reprinted by permission of Random Hosue, Inc.

second. The fission fragments are usually nuclei of lighter elements, such as strontium or cesium. The fission heat is absorbed by the reactor coolant and contributes to the generation of steam. In a light-water reactor, the coolant also acts as a "moderator" by slowing down fast neutrons emitted during fission. Given any volume of material, *slower* neutrons are more likely to split nearby fissionable nuclei than *fast* neutrons; thus more nuclei will split when *slow* neutrons are more abundant, and more energy will be released.

If the uranium nucleus emits free neutrons, these speed away and may similarly strike other nearby uranium nuclei. A fission chain reaction is self-sustaining when the number of neutrons released in a given time equals or exceeds the number of neutrons that are absorbed by nonfissioning material or that escape from the system. We say that a nuclear reactor is "critical" or has "reached criticality" when it is sustaining a chain reaction.

When an atomic bomb is exploded, this chain reaction occurs more rapidly. For example, one fissioned nucleus will emit two neutrons, which are absorbed by two other uranium nuclei, which release *four* neutrons, and so on. At the same time, a gun or implosion mechanism contains or increases the density of the fissionable material to sustain a critical mass. The resulting explosion releases a tremendous amount of energy in a very short time and reaches temperatures in the millions of degrees. In a power reactor, the reaction is controlled to the desired energy release rate.

To control the multiplication of neutrons in a reactor core, control rods containing a neutron absorber such as boron are moved in and out. These control rods absorb some of the free neutrons emitted during fission. When *all* the control rods are inserted, the boron near the fissioning uranium absorbs enough of these free neutrons so that the rate of fission levels off. If enough control rods are withdrawn quickly, the fission reactions start multiplying very rapidly and may "run away." If the resulting energy release increases beyond the safe point, the fuel may be damaged, the steam pressure will increase, and possibly the steam system may suffer overpressurization. Such an event is called a "transient accident."

## Questions for Discussion

1. What are the audience and purpose of the process description above? Why does this description begin with a contrast of nuclear plants to fossil-fuel plants?

2. What are the basic differences between BWR, PWR, and LMFBR plants? Why does the writer describe each type? Does he use device description in doing so?

3. What are the principles involved in the nuclear plant process? Do these principles vary between the BWR, PWR, and LMFBR? How?

4. What is the main difference in the nuclear fission process between an atomic bomb and a nuclear plant?

5. What is a "transient accident"?

6. What specific purposes do the graphic aids serve in this process description?

7. Imagine that you are explaining the nuclear plant process to a group of school children. Choose one type of plant and explain the plant operation so that this audience would understand this complicated process.

# Exercises and Assignments

1. **Problem:** When we describe a process for an audience, often we can observe the device or system in operation. When we record our observations in a process description, we must represent action by words. If we are relying on our own sense impressions, we should make sure that our observations are objective and accurate. One way we can check a process description is to compare it with someone else's observation of the same process.

Assignment:

1. Choose a device with three to five moving parts, for example, an egg beater, an office-size pencil sharpener, or a manual can opener.
2. Pair up with another student and observe the device in operation.
3. Each one of you write a process description of the device.
4. Compare your description with your fellow student's. Analyze the similarities and the differences.

2. **Problem:** A process diagram often accompanies a description of a device in operation. This diagram shows the parts of the device and their interaction. Arrows within the diagram indicate the direction of movement or action. The verbs we choose to use within a process description should help an audience visualize the precise action depicted in such a diagram.

Assignment:

Examine Figure 9-8. The device in the diagram is a clothes dryer. Write a process description of the dryer in operation. Choose verbs that represent the action in the process exactly. Be sure to study the direction of the arrows in the diagram. Designate an appropriate audience and purpose for your description.

Assignment:

Examine Figure 9-9. The device in the diagram is a slide projector. Write a process description of the projector in operation. Choose verbs carefully and study the direction of the arrows in the diagram. Designate an appropriate audience and purpose for your description.

3. **Problem:** As we studied in this chapter, the amount and type of detail we include in, and even where we begin and end, a process description depend on our audience and purpose. For technical, lateral audiences we usually include many technical details about each step in the process. For managers, some employees, the public, and potential customers, we might include only the major steps in the process.

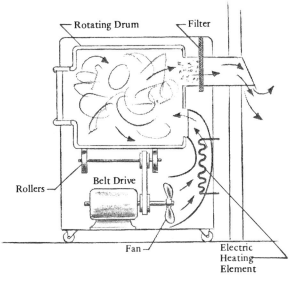

*Figure 9-8.* From *How Does It Work?* by Richard M. Koff and illustrations by Richard E. Rooman. Copyright © 1961 by Richard M. Koff. Reproduced by permission of Doubleday & Company, Inc.

## Assignment:

**1.** Choose a device with at least five moving parts.
**2.** Write a process description of the device for a technical audience. Place yourself and your audience in a hypothetical situation or role.

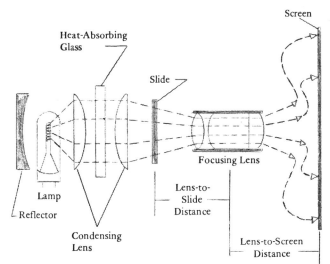

*Figure 9-9.* From *How Does It Work?* by Richard M. Koff and illustrations by Richard E. Rooman. Copyright © 1961 by Richard M. Koff. Reproduced by permission of Doubleday & Company, Inc.

3. Write a process description of the same device for a "general" audience. This audience may be supervisors or managers, potential customers, employees, or the public. The type and amount of detail you include depend on the audience you choose. Make sure, however, that your process description is different from the one you did in No. 2 of this assignment.

4. Write a process description of the same device for a child. Be sure to compare the steps within the process with actions that will be familiar to your audience.

4. **Problem:**  Assume that you are an engineer or technologist for a growing manufacturing company. You have developed a new device that you would like your company to produce and market. This device can be either an improvement on an old device you make now or a device your company has never produced. You are to write an upward-directed communication to your supervisor which recommends that your company produce and market the device.

### Assignment:

1. Choose two types of devices. These devices can be common household items, for example, two types of can openers, two types of coffee makers, or two types of hair dryers.

2. Write the upward-directed communication in which you recommend one of these devices. For example, you might recommend one type of can opener rather than the other. Again assume that your company already produces one type and that you are recommending they produce the other instead, or assume that you want your company to enter this market and produce one type rather than the other. In the second case, the other type of device might be your competitor's product. Within your communication, be sure to describe and compare and contrast the operation of both types of devices.

# 10

# Instructions:
## Directing How to Use or Repair a Device

*Selecting detail and sequence in instructions.*
*Addressing the audience directly.*
*Writing operating and troubleshooting instructions.*
*Including notes, cautions, and warnings in instructions.*
*Supporting instructions with graphic aids.*
*Reading and analyzing instructions written by others.*

Let us now turn to another type of technical narrative: instructions. While in process description our purpose is to help an audience visualize an operation, in instructions we tell an audience how to *participate* in an operation. Instructions are detailed directions that enable the audience to carry out a specific task. Because this audience may actually complete the operation while reading the instructions, we address its members directly. Instructions must be accurate, complete, and safe. In instructions we tell how to install, assemble, operate, maintain, or repair a device. Although sometimes in instructions we tell an audience how to complete an action when no equipment is involved, in this chapter we concentrate on instructions for using a device. The general consumer as well as the expert needs instructions on how to use equipment properly.

Perhaps some of the first technical instructions ever recorded were directions for conducting experiments. For example, in 1755, Benjamin Franklin gave instructions for conducting experiments with electricity. Franklin first stated the principles of electricity and then gave not only directions for the experiments but also the cause of the predicted results:

### Preparation

Fix a tassel or fifteen or twenty threads, three inches long, at one end of a tin prime-conductor (mine is about five feet long, and four inches in diameter) supported by silk lines.

Let the threads be a little damp, but not wet.

### Experiment I

*Pass an excited glass tube near the other end of the prime-conductor, so as to give it some sparks, and the threads will diverge.*

Because each thread, as well as the prime-conductor, has acquired an electric atmosphere, which repels and is repelled by the atmospheres of the other threads: if those several atmospheres would readily mix, the threads might unite, and hang in the middle of one atmosphere, common to them all.

*Rub the tube afresh, and approach the prime-conductor therewith, crossways, near that end, but not nigh enough to give sparks; and the threads will diverge a little more.*

Because the atmosphere of the prime-conductor is pressed by the atmosphere of the excited tube, and driven towards the end where the threads are, by which each thread acquires more atmosphere.

*Withdraw the tube, and they will close as much.*

They close as much, and no more; because the atmosphere of the glass tube not having mixed with the atmosphere of the prime-conductor, is withdrawn intire, having made no addition to, or diminution from it. . . .

Franklin's instructions are so simple and complete that anyone could conduct the experiment. In instructions we speak directly to our audience so that they can duplicate easily the actions we present; we *narrate* an operation in which they participate.

# Sequence in Instructions

Generally, because of the basic purpose of instructions, sequence must be chronological. The audience follows the directions from the first to the last step. If the steps are out of order, the audience will not complete the action successfully. The only time we might not follow chronological order in instructions would be if we presented alternative steps. For example, if some of the audience had a different model of device, they might skip a step, or, when "troubleshooting," if one method did not solve the problem, the audience might repeat a step. We see examples of troubleshooting later. Basically, however, we use chronological order in instructions.

# Selection of Detail in Instructions

We must choose when to begin and end our instructions and how much and what kind of detail to include. These decisions depend on the audiences and purposes of our instructions. We look at common audiences of instructions next.

## Instructions for the General Consumer

*Using a Specific Device.* Usually accompanying every new appliance or piece of equipment we buy or sell is an instruction sheet or booklet. These instructions tell how to unpack, assemble, and use the device as well as what to do should the device not operate properly. When we write instructions for the general consumer, we must write simply and clearly and assume no prior knowledge of the device on the part of the audience. We start with the earliest possible action and include each action no matter how simple.

For example, in the following instructions, the writer teaches the audience how to operate an instamatic camera properly and how to take good pictures. Notice how clear and simple the instructions are:

> 1. Make sure there is film in the camera, and that if there is, there are exposures left on the roll. If there is a film cartridge in there, but no pictures left, remove the old cartridge and put in a new one. If there is no film in there, put in a new roll.
> 2. Check to make sure that you have operated the film advance until it locked after your took your last picture. This locking feature of the film advance is to prevent you from losing a picture by double exposing a frame. You will not be able to push the shutter release down if the film advance has not locked.
>
> *Caution:* Do not try to force the shutter release down if it will not go.
>
> Check to see if the film advance has been operated until locked.
> 3. Stand with the sun behind you if you will be taking the picture in bright or hazy sunlight.

4. Insert a used flashcube into the flashcube socket if you will be taking the picture in cloudy-bright conditions when there are no distinct shadows.

5. Insert a flashcube with a fresh bulb facing forward in the socket if you will be taking the picture in the shade. Remember that for flash pictures such as these there must be between 5 to 9 feet between you and your subject.

6. Always keep at least 5 feet away from your subject when taking pictures so that they will be sharp.

7. Hold the camera horizontally or vertically (with the lens at the top and shutter release at the bottom) depending on the subject and which way it fits best in the viewfinder.

8. View your subject with your eye close to the viewfinder opening.

9. Frame your subject in the viewfinder.

10. Hold the camera steady, keeping your hands and the wrist strap behind the front edge of the camera. Gently squeeze the shutter release. Do not jerk the camera by punching the shutter release or your pictures will be blurred.

11. Push the film advance until it locks.

*Paula Huganine*

Since in instructions for the general consumer, we must assume that the audience has never operated the device before, the steps in our instructions must be detailed and the action clearly described. As in the instructions above, we begin with the earliest possible step and end with the latest.

***Following a General Operation.***    In some cases we must give instructions on how to carry out a general operation that our audience adapts to particular situations. In this case, we must introduce our instructions by telling how they apply in various circumstances or with various materials. This introduction helps our audience gather materials or prepare a work area first, so that we can describe the basic action steps simply and concisely.

For example, in the following instructions, the writer gives directions for soldering metal. Since the overall procedure applies to small or large jobs; wire, bar, or paste solder mixtures; and rosin, acid, or galvanized fluxes, the writer gives all possible variations before the basic instructions. Because so much explanation appears in the introduction, the instructions are simple and easy to follow:

Soldering is an easy way to mend metals permanently. Small soldering jobs are best done with an electric soldering iron of the conventional type or of the faster gun type. Large jobs are done with a torch (a conventional gasoline blowtorch or the more convenient propane torch). The tips of soldering irons must be kept smooth, clean, and tinned (coated with a thin layer of solder). Torch tips must also be kept clean and undamaged if they are to give a steady blue flame of the proper size and shape.

Solder is a mixture of tin and lead. It is available in the form of wire, bars, or paste. Wire solder may be solid solder or contain a core of acid or rosin flux. Bar solder is solid. Paste solder comes mixed with acid flux.

Whenever solid solder is used, a flux which cleans the metal must first be applied to the metal. This is available as a liquid or paste. Rosin flux is used only in making electrical connections and on terne metal. Acid flux is used for all other work. Galvanized iron and stainless steel require especially strong acid fluxes.

Aluminum is soldered without fluxing, but a special solder called Chemalloy is required.

Here is the step-by-step procedure for soldering:

1. If the object to be soldered contains a liquid or gas, empty it completely.

2. Clean the metal with steel wool until it shines brightly.

3. Apply flux to the cleaned surface if the solder does not contain flux.

4. Heat the metal until it is hot enough to melt the solder applied to it. Except in the case of lead and pewter, the solder should always be melted by the metal being soldered—not by the soldering iron or torch. If the metal is not heated sufficiently, the melted solder looks dull, granular, and rough. It should flow and form a smooth, bright film.

5. To join two pieces of metal (except in plumbing work), it is advisable to tin (flow a film of solder on) each piece separately. Then place tinned surface to tinned surface and apply heat until the solder melts. Remove heat and hold the metal pieces together with a screwdriver, pliers, etc. until the solder hardens.

6. To solder copper pipe into a plumbing fitting, clean both pipe and fitting and brush on an acid flux. Insert pipe into fitting and apply heat to the joint. Then touch solid solder to the joint. It will be drawn by capillarity into the joint and seal it.

7. When acid flux is used, wash off the residue thoroughly with water after the metal has cooled.

The more we "set-up" our instructions in an introduction, the easier they will be to follow. Outward-directed instructions for the general consumer must be clear and simple and, if necessary, well introduced.

# Instructions for a Technical Audience

*Using a Specific Device.* In many instructions we address a specialized audience, an experienced employee, or a peer. (Seldom do we send instructions upward to a supervisor or manager.) Manuals or instruction sheets and booklets usually accompany complex devices and equipment used by engineers and technicians. In these we can often assume that our audience has a certain education, understands a specialized vocabulary, and may have used a similar device before.

For example, the following instructions appeared in a manual which contained unpacking, assembly, installation, and operating instructions for a valve body. In the excerpt the writer tells how to "make up" the stem connector of the valve. Although the writing is as clear as that in the examples directed toward the consumer, the vocabulary is specialized and the steps complex. The audience has to study each step while comparing actual parts on the device with the parts described. The audience must complete the action represented in the illustrations slowly and carefully:

1. Move the valve plug to the closed position.
2. Screw the locknuts onto the stem and set the travel indicator disc on these nuts with the "cupped" portion of the disc facing downward.
3. Move the valve plug stem up the required travel and attach the stem connector, making sure that there is full engagement of the actuator stem threads. Install the two cap screws in the stem connector, but tighten them only slightly at this time. Tighten the stem locknuts slightly to position the travel indicator disc against the bottom of the stem connector.
4. The travel indicator should show the valve to be wide open with no pressure on the diaphragm. If it does not, loosen the screws that hold the travel indicator scale and shift the scale to the required position.
5. Apply and vary the pressure to the diaphragm case and observe the valve travel. Make sure the valve plug seats on the seat ring.

CAUTION

On bellows seal bonnets, the stem must not be rotated, or damage to the bellows will result.

5.1 If the travel on units with plain or extension bonnets is not correct, it can be changed by screwing the valve plug stem into or out of the stem connector. Turn the stem by using a wrench on the locknuts; never use a pipe wrench or pliers on the stem itself and never turn the plug while it is on the seat.
5.2 To adjust travel on units with bellows seal bonnets, shut off supply pressure and disconnect the pressure line. Loosen the yoke locknut (remove the actuator-to-bonnet bolts and nuts on units with five-inch yoke bosses) and rotate the entire actuator to reposition the stem connector on the valve plug stem. If it is not possible to rotate the actuator, remove the stem connector, adjust actuator signal to reposition the actuator stem, and install the stem connector. When travel is set properly, tighten the cap screws in the stem connector and lock the stem locknuts against the stem connector.

The writer assumes that the audience is capable of completing the actions and identifying the parts of the device. Although many steps include more than one action and call for the audience to judge whether the action has been successful, the writing is still simple and concise.

***Following a General Operation.*** With a technical audience as with the general consumer, sometimes we give instructions that do not involve a specific device, or we ask the audience to adapt an overall operation to a particular situation. We assume a certain degree of expertise on the part of our audience and use detail to go beyond the basic steps we might use in directing a general audience.

For example, in the following instructions, the writer tells physicians how to treat patients poisoned by a certain pesticide. The writer uses a specialized vocabulary, includes much detail, and assumes that the audience can select the steps that pertain to particular situations. Although each step in the instructions is complicated, the writer is still clear and concise:

> 1. Establish CLEAR AIRWAY and TISSUE OXYGENATION by aspiration of secretions, and if necessary, by assisted pulmonary ventilation with oxygen.
>
> 2. CONTROL CONVULSIONS. The anticonvulsant of choice is DIAZEPAM (VALIUM). Adult dosage, including children over 6 years of age or 23 kg in weight: inject 5–10 mgm (1–2 ml) slowly intravenously (no faster than one ml per minute), or give total dose intramuscularly (deep). Repeat in 2–4 hours if needed.
>
> Dosage for children under 6 years or 23 kg in weight: inject 0.1–0.2 mgm/kg (0.02–0.04 ml/kg) slowly intravenously (no faster than one-half total dose/minute), or give total dose intramuscularly (deep). Repeat in 2–4 hours if needed.

> CAUTION: Administer intravenous injection slowly to avoid irritation of the vein, occasional hypotension, and respiratory depression.

> Because of a greater tendency to cause respiratory depression, BARBITURATES are probably of less value than DIAZEPAM. One used successfully in the past is PENTOBARBITAL (NEMBUTAL). Maximum safe dose: 5 mgm/kg body weight, or 0.20 ml/kg body weight, using the usual 2.5% solution.

> If possible, inject solution intravenously, at a rate not exceeding one ml/minute until convulsions are controlled. If intravenous administration is not possible, give total dose rectally, not exceeding 5 mgm/kg body weight (0.2 ml/kg of 2.5% solution).

> CAUTION: Be prepared to assist pulmonary ventilation mechanically if respiration is depressed.

The writer capitalizes important actions and medications so that even this specialized audience can follow easily the instructions.

Instructions then are usually directed downward, outward, or laterally. Although we seldom command peers, we can teach them how to improve their jobs or how to operate a device in our own specialty. When we instruct a general consumer, we give complete and simple directions on how to carry out each step in the action. We often need to write for an audience with as low as a third-grade reading level. However, when we write instructions for a technical audience, we can use a specialized vocabulary and assume that our audience may be experienced in similar operations. We select the type and amount of detail to include based on our audience and purpose. Let us look further at two special purposes of instructions

# Operating and Troubleshooting Instructions

We also select the type and amount of detail to include in instructions according to our overall purpose: operating or troubleshooting. Operating instructions are usually less complicated and require less study than do troubleshooting instructions. Operating instructions also follow a basic chronological order, whereas troubleshooting directions may offer the audience several alternative steps.

## Operating Instructions

Most of the instructions that we have looked at so far are operating instructions. For technical audiences, operating instructions are usually preceded by a device description and a process description. Again operating instructions are ordered chronologically, and the steps refer to one basic action.

For example, the following are operating instructions for a transmitter-receiver. The writer includes basically one action per step which the technical audience completes before going on to the next:

### 3.5 Operation

Operating with headphones connected is the same as with a HANDSET or using the front panel speaker. Set RECEIVE AUDIO gain control for a comfortable listening level.

NOTE

*If headphones are used, the received audio signal is disconnected from the handset.*

### 3.5.1 Receiver Adjustment

The following procedure should be used to set the operating controls for receiving.

a. Warm up set as explained in paragraph 3.4 [not included here].
b. Select frequency as explained in paragraph 3.3 [not included here].
c. Set FUNCTION switch to desired mode.
d. Turn SPEAKER SWITCH to ON position.
e. Set RF GAIN control to maximum clockwise position. Set TRANSMIT AUDIO control fully clockwise to TUNE detent.
f. Set RECEIVE AUDIO gain control for comfortable listening level.
g. Set the PRESELECTOR control somewhat below the lower limit of MHz band to be used and tune up into band for first noise peak.

NOTE

*It is possible to tune the PRESELECTOR to spurious signals at other points on the dial, particularly above the selected frequency. By starting below this frequency and tuning upward to it, the chances of tuning to a spurious signal are minimized. Check PRESELECTOR dial indication to ensure it is peaked near the proper frequency, allowing for calibration errors.*

h. To tune to a transmitted signal rotate the TRANSMIT AUDIO control fully counterclockwise and tune PRESELECTOR for peak indication on meter. If the signal appears to be off-frequency, use vernier (VFO) tuning as described in paragraph 3.3 [not included here].
i. Readjust RECEIVE AUDIO gain control if necessary. If ignition or other pulse type noise becomes objectionable, turn NOISE LIMITER ON. . . .

Operating instructions direct an audience on how to use the device with success. To write thorough operating instructions, we must anticipate all the questions that our audience might have; we have to use our words to take the place of an on-the-spot teacher.

## Troubleshooting Instructions

In troubleshooting instructions, we help our audience analyze what is wrong with a device and repair it. Because the steps in trouble-shooting instructions indicate each stage of problem analysis, the audience often has a choice of action. In troubleshooting instructions, we have to anticipate and visualize all that might possibly go wrong with the device and help our audience diagnose the problem and then solve it. We have to use our words to take the place of an on-the-spot advisor.

For example, in the following troubleshooting instructions for the same device addressed in the preceding operating instructions, the writers help the audience find and solve problems with the transmitter:

### 6.6.2 Transmitter Failure

a. Remove top and bottom covers from transceiver. Remove ALC assembly. . . .

b. Connect dummy load to ANTENNA connector J4.

c. Apply power, and set transmitter for CW output at 2.555 MHz with TRANSMIT AUDIO control set at mid point. Key the transceiver. CW sidetone should be present at the speaker. Unkey transmitter and tune PRESELECTOR control for maximum noise output from speaker. Perform steps (d) through (g) with transmitter keyed. (Leave transmitter keyed only as long as is necessary for each step since fan is disconnected.)

d. Disconnect P11 at translator and connect P11 to signal generator set for 400 MV output at 2.555 MHz. Tune PRESELECTOR and PA TUNE controls for maximum indication; on transceiver output 100W watts or more, proceed to step g.

e. Connect RF VTVM to pin 11 of K1 located on the RF amplifier assembly. . . . If indication is 400 MV, proceed to step f. If no indication is present, check K1 and the coaxial cable conected from K1 pin 9 to J50.

f. Connect RF VTVM to output of RF Amplifier at terminal E159. . . . Slowly tune PRESELECTOR for peak indicated of 35 volts or more of VTVM. If no or low output is present at E159, proceed to paragraph 6.9. If output is normal, proceed to paragraph 6.8 [these paragraphs appeared elsewhere in manual].

g. Unplug P10 at translator assembly, and connect it to RF VTVM. Input signal level at P10 from upper sideband filter should be from 15 to 30 MV. A 10 MV signal will result in an RF output from PA section of at least 50 watts. If normal signal is present, problem is probably in the Translator T/R or R/T diode switching circuits. Check the voltages at Translator R/T terminal E62 (0 VDC) and T/R terminal E63 . . . (12 VDC). If no or low signal is present at P10, problem is in stages preceding Translator. Reconnect P10.

The audience of these troubleshooting instructions would study each action and the results of that action before going on to the next step. When writing troubleshooting instructions, we must anticipate everything that may go wrong with a device as we help our audience analyze and solve the problem.

Therefore, operating instructions are usually less detailed, read more quickly, and direct step-by-step action, while troubleshooting instructions demand more analysis and participation on the part of the audience and contain choices of action. These special purposes of instructions determine the amount and type of detail to include.

## Alerting an Audience to Danger

Instructions must not only be complete, concise, and readable, but they must also be safe. We must alert our audience to any potential danger to themselves or to the equipment. In instructions, we use a certain code to alert the audience. This code includes "warnings," "cautions," and "notes."

Warnings alert the audience to potential personal danger. Two examples of warnings follow:

> WARNING

To avoid personal injury and damage to the process system, isolate the control valve from the system and release all pressure from the body and actuator before disassembling.

> WARNING

Overpressuring any of the system components could cause personal injury, equipment damage, and fire or explosion hazard due to venting of hazardous supply pressure medium. To avoid such injury or damage, provide suitable overpressure protection devices to protect all system components from overpressure.

Cautions alert the audience to potential equipment damage. Two examples of cautions follow:

> CAUTION

Never use an old stem with a new valve plug. The use of an old stem requires drilling a new groove pin hole through the stem, thereby weakening the stem.

> CAUTION

The exposed portion of the cage provides a guiding surface that must not be damaged during disassembly or maintenance. If the cage is stuck in the body, use a rubber mallet to strike the exposed portion at several points around the circumference.

Notes give the audience important information on model differences, choices in action, and possible problems in operation. Two examples of notes follow:

Note

If it is not possible to provide a temperature equal to the upper-range limit of the temperature bulb, use any temperature that is available within the range. Then, rotate the temperature setting knob to the setting that corresponds to the temperature of the temperature bulb.

Note

Spiral wound gasket boltup characteristics are such that tightening of one bolt may loosen an adjacent bolt. This will occur on subsequent tightening of all the bolts until the bonnet-to-body seal is made. This requires several trials on each bolt until the nut does not turn at the given torque.

The words "warning," "caution," and "note" are usually centered on the line. Warnings and cautions should precede the step to which they refer. The words "warning" and "caution" are usually capitalized and boxed to draw attention to them. The most effective warnings and cautions not only alert the audience to the importance of following a step closely, but also tell exactly what will happen if the audience disregards the alert; explicit reasons motivate the audience to be cautious. In instructions, we must make sure that our audience can follow our directions safely.

## Style in Instructions

Since instructions address the audience directly, we use the imperative voice when indicating action. Rather than saying,

You should loosen the wheel nuts

or

The wheel nuts are now loosened

we say

Loosen the wheel nuts.

We should also use the precise verb in instructions. We say "loosen" rather than "detach" or "unfasten" if we want the audience to unscrew the wheel nuts somewhat but not take them off.

Generally the steps in instructions are numbered and indented in paragraph or list form. If possible, steps should be short and refer to one action, especially in instructions that address the consumer. We can subdivide steps if they are too long (step 3; step 3.1, 3.2, 3.3, etc.).

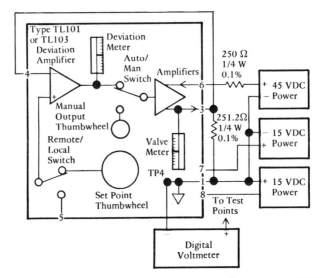

*Figure 10-1.*    Troubleshooting setup. Courtesy Harris Corporation, RF Division.

## Graphic Aids in Instructions

Graphic aids are often essential in instructions. As with device description, graphic aids show audiences on paper what they should be seeing on the device in front of them. Graphic aids also simplify instructions; we do not have to locate the part for the audience each time we mention it.

Graphs, tables, and diagrams are all helpful to the audience. For example, in Figure 10-1, the parts of the device are indicated that the audience must operate while "troubleshooting." Notice that

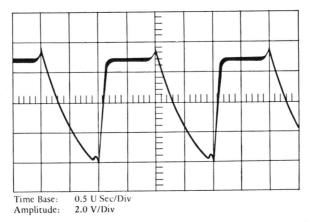

Time Base:    0.5 U Sec/Div
Amplitude:    2.0 V/Div

*Figure 10-2.*    Collector waveform. Reconnect oscilloscope to collector of 500 kHz multivibrator stage Q3 (or top of R7). The observed waveform should resemble that shown in figure. Courtesy Harris Corporation, RF Division.

TABLE 10-1. Troubleshooting a Frequency Controller

| Symptoms | Possible Causes | Check |
|---|---|---|
| No indicators or display | (a) No +13.2 Vdc from RF-230 | (a) Check interface between RF-230 and RF-252. |
| | (b) Bad +10 Vdc Power Supply | (b) Check +10 Vdc power supply on Main Control PWB A2A1. |
| Any signal frequency or channel display fails to illuminate | (a) Bad Display Select Circuitry | (a) Check A2A1U1. |
| | (b) Bad Darlington Transistor | (b) Check A1A2Q2 through A1A2Q7. |
| | (c) Bad Display Segment | (c) Check A1A2DS1 through A1A2DS6. |

the parts are labeled by their entire names rather than by a code so that the audience can better match the instructions, the graphic aids, and the actual device.

Troubleshooting tables are often included in manuals. These tables, such as Table 10-1, help the audience compare and analyze a variety of "symptoms," causes, and solutions when a device does not work properly.

Graphs are also useful in instructions. Figure 10-2 shows the pattern that the audience would see on an oscilloscope if the device were working properly. Line graphs sometimes show movement better than diagrams and tables.

# Summary

Instructions are a special type of technical narrative in which the audience participates directly and fully in the action. In instructions, we give clear, concise, complete and safe step-by-step directions on how to operate or fix a device. We usually write instructions for the general consumer or for a specialized audience in downward or lateral communications. While we generally order instructions chronologically, we choose the amount and type of detail and the beginning and ending step according to our audience and purpose. Nonspecialized audiences need more introductory material in instructions.

Two special purposes of instructions are operating and troubleshooting. In troubleshooting instructions, we often indicate alternative steps and ask our audience to analyze carefully each option. We include warnings, cautions, and notes to make both operating and troubleshooting instructions safe to follow. Instructions must be readable and clear and address the audience directly. Graphic aids, such as diagrams, tables, and graphs, help audiences visualize the device they must operate and the action they must take.

# Readings
# to Analyze and Discuss

## I
## A

### 5.8.1 Receiver Protector Test

To determine if the receiver protector assembly is operating, perform the procedures outlined below.

a. Energize the transceiver and tune to receive at 2.0 MHz. Set FUNCTION switch at USB.
b. Rotate TRANSMIT AUDIO control fully counterclockwise.
c. Connect an RF signal generator, such as a HP-606A, to the ANTENNA connector on the rear panel using a coaxial cable. Set the RF signal generator for 1 μ volt output at 2 MHz.

---
CAUTION
---

*Do not key transceiver with RF signal generator connected to ANTENNA connector.*

d. While monitoring the received RF signal generator signal rotate the TRANSMIT AUDIO control fully clockwise to the TUNE detent. Observe that the received signal is no longer present.
e. Set the FUNCTION switch at OFF. Disconnect the RF signal generator.

## B

### 6.6.3 Receiver Failure

a. Remove top and bottom covers from transceiver. Connect dummy load to ANTENNA connector J4.
b. Apply power and tune transmitter for 100W CW output at 2.555 MHz.
c. Disconnect microphone or CW key from transmitter. Connect a signal generator set for a 10 MV, 2.555 MHz signal to ANTENNA connector J4 in place of dummy load. Set RF GAIN fully clockwise.
d. Unplug P11 at Translator, and connect it to RF VTVM. Vary output frequency of signal generator above and below 2.555 MHz while observing RF Amplifier output indication on RF VTVM. A peak indication of at least 50 MV should be obtained. If not, problem is in antenna transfer relay K4, Receiver Protector assembly, RF Amplifier transfer relay K1, or cabling. If this signal is present, problem is in receive circuits following RF Amplifier. Reconnect P11.

## Questions for Discussion

1. Contrast the audiences and purposes of the instructions above. Are they operating or troubleshooting instructions?

2. What differences do you see in the style and format of the instructions? How do you account for these differences?

**194**

3. What types of graphic aids could be used to support the instructions?

4. Describe the detail and sequence in both sets of instructions.

# II
# Fundamentals of *ROSCOE*

## Modifying Programs Before They Are Saved

If you have discovered some errors after listing your program, you can make corrections using one of several methods. If you have found an error in one of your lines, you can correct it by simply typing that line number followed by a space, and then retyping the correct line of information and hitting the return key.

Or, you can use the EDIT or E command. You may edit a line by typing EDIT or the abbreviation E, a space, a slash, a unique string of characters containing the error, another slash, the correct string of characters, another slash, and then the line number in which the error appears. For example, if you have a line which reads:

0050 WRITE (6,2000) BAS,AA,NWW,NE

and you want it to read:

0050 WRITE (6,2000) BAS,AE,NWW,NE

you can correct it using the EDIT command by typing

E /AA/AE/50

Note that if you type only:

E /A/E/50

all of the A's in that line will be changed to E's. Similarly, if you do not type the line number in which the error appears, all of the A's in your entire program will be changed to E's.

If you wish to add a new line to your program, simply type the new line, being sure to give it a line number which falls between the two lines where you want the line to be inserted. For example, if you want to insert a new line between the following two lines:

0020 AVG = TOTAL/N

0030 GO TO 300

you can assign the new line any line number between 0020 and 0030 (0021 to 0029). When you list your program again, the new line will appear in the proper sequence.

If you wish to get rid of a line, simply type the command DELETE (no abbreviation), followed by a space and the line number of the line you wish to erase.

### NOTE

If you do not type the line number after the word DELETE, you will erase your entire program.

## Storing Programs

*the SAVE command, the LIBRARY command*

If you wish to store your program in your ROSCOE library so that you can use it at a later date, you can file it using the SAVE or S command. After you have listed your program and corrected all the errors, simply type the command SAVE or the abbreviation S, followed by a space and any string of characters you wish, beginning with a letter and not exceeding eight characters. This string of characters has now become the name of the data set in your ROSCOE library which contains this program.

To review the list of data sets contained in your library, type the command LIBRARY and hit return. A list of all the data sets you have saved will then print along with the date saved, the most recent revision date, and the length of each data set.

CAUTION

Be sure you have saved your program before listing your library. Otherwise, the LIBRARY command will destroy your program and replace it with the previously discussed list of data sets.

## *Questions for Discussion*

1. What are the audience and purpose of the instructions above?

2. What is different about these instructions in contrast to those we looked at in this chapter?

3. What is the purpose of the examples in the instructions?

4. Rewrite the instructions in the imperative voice, with numbered, indented steps, one action per step. What do you gain or lose when you rewrite the original in this way?

# *Exercises and Assignments*

**1. Problem:** Pair up with another student. Choose an operation or action that you both know well.

Assignment:

1. Without consulting the other student, each write your own set of instructions. You may choose something simple such as changing a tire on a car. Decide on a specific audience and purpose.
2. Exchange your instructions with the other student. Describe and account for the differences in your instructions.
3. Write a device and process description to precede your instructions.

**2. Problem:** Instructions that are directed toward a general audience such as the consumer must be precise, complete, and easy to follow. We must assume that the audience might have never seen the device before.

Assignment:

Write a set of instructions in which you address the general consumer. Assume that your audience has never used your device before. For a challenge, you might assume that your audience consists of children; your device might be a child's toy or model. Or assume that your audience is at a disadvantage in some way; for example, your audience might be blind, or read English as a second language.

**3. Problem:** In operating instructions, we instruct an audience on how to use the device properly and safely. In troubleshooting instructions, we help an audience find the causes for device failure and fix the device. Study carefully a device with which you are familiar.

Assignment:

1. Write a set of operating instructions for the device. Be sure to use warnings, cautions, and notes correctly. Designate an audience and purpose for your instructions.
2. Write a set of troubleshooting directions for the device. Be sure to give all possible reasons for device failure and all possible ways to fix the device. Hint: Choose a simple device such as a stapler. You might instruct your audience on how to load and operate the stapler for No. 1 of this assignment and help the audience fix the stapler should the staples not bend properly or jam within the stapler for No. 2.
3. Write a device and process description to precede your instructions.

# 11

# Procedures:
## Standardizing Operations
# and Specifications:
## Prescribing Criteria

*Choosing a sequence and amount of detail in procedures.*
*Indicating alternative approaches in procedures.*
*Clarifying options and requirements in procedures.*
*Writing prescriptive and performance specifications.*
*Sequencing specifications.*
*Expressing alternatives in specifications.*
*Deriving specifications.*
*Reading and analyzing procedures and specifications*
*   written by others.*

In this chapter, we study two more types of technical narrative: procedures and specifications. In procedures, we describe in general terms how to handle an operation in which many people may be involved. In specifications, we prescribe the particular materials, methods, and objectives of projects or proposed operations to those who will undertake them. Procedures fall on one side and specifications the other side of instructions in terms of type and amount of detail. However, both procedures and specifications address expert audiences.

In procedures, we address an audience already familiar with our subject and experienced in performing some or all of the operation we describe. We may relate improvements or changes in existing procedures, we may clarify a procedure for the many people involved, or we may formalize or standardize a procedure that has existed for awhile. In any case, we assume that the people involved in the procedure have been or will be instructed on how to carry out their specific jobs in some other communication. We use procedure to show how many people in the audience cooperate to achieve a certain goal. Since we describe a standard operation and assume that our audience will adjust the procedure to each particular situation, procedures are seldom as detailed as instructions.

Thus, in procedures, we describe an overall operation that usually involves more than one person; for example:

> When a customer first comes into the Complaint Department, the clerk completes the regular complaint form which calls for specific information such as date purchased and cost as well as comments on product performance and service. The blue copy of the form is placed in departmental files, the pink copy sent to the department manager where the product was purchased, the green copy sent to Personnel, and the white copy retained by the customer. The blue, green, and pink copies are kept for ten years and then destroyed. If a customer complains specifically about service and the purchase was for $150 or more, a Personnel Department representative interviews the customer. If the customer complains about the product and the purchase was for $150 or more, the specific department manager interviews the customer. The content and results of the interviews are summarized in monthly reports. If the customer has filed a previous complaint either about the same product or service or about a prior one, the Complaint Department manager interviews the customer after the clerk has completed the form.

In the overall goal of satisfying the customer, the Complaint Department clerk, a department manager, the Personnel representative and manager, and the Complaint Department manager all participate.

Because in specifications, we give the particulars or requirements of projected work, specifications are more detailed than are procedures and even instructions. Specifications are, as the name implies, "specific." Usually accompanied by land maps or blueprints, specifications spell out exactly what methods (or actions) and materials must be used on a job. Specifications are most common in the construction industry, in which owners contract firms to build structures. Engineers, architects, and designers draw up the plans and recommend specifications that owners require contractors to follow. Specifications are often legal documents and must incorporate codes and standards set up by agencies and associations in the field. For example, a contractor building a highway might follow the requirements of the Portland Cement Association, the American Society for Testing and Materials (ASTM), and the Standard Specifications for Welding Structural Steel along with many others, as well as the specifications drawn up by the engineers for that particular project. Moreover, the specifications writer for the highway project might follow the format guidelines of the Construction Specifications Institute.

Thus, in specifications, we set the requirements or provisions for a projected job; for example:

**Requirements for Form 19.7 (Complaint) Reorder for H. J. Shipley, Inc.**

Section 1. Form 19.7 shall be printed on 8½ by 11 inch pads, 1000 sheets each. Tops of sheets shall be gummed with Apoxic 179. Sheets shall be perforated at the top, one half inch below the gummed margin. Sheets shall alternate in color: white, blue, pink, and green. The sheets shall be printed in black ink, one side of page. Removable carbon sheets shall be placed between every sheet.

1.1 The printer will contract with a paper supplier whose bid will be reviewed by Shipley, Inc.

1.2 The printer will be responsible for delivery of order.

In specifications, we prescribe the exact materials, methods, and final products for a job.

# Procedures

Generally the members of an audience of a procedure are familiar with their jobs, share a common goal, and all participate in the procedure in some way. Audiences or procedures may need to understand company procedure before they engage a company's services, may need to have an old procedure clarified or a new procedure justified, or may want to adopt a procedure. Because a procedure is a type of narrative, we need to decide on sequence and the amount and type of detail.

# Sequence in Procedures

If the exact order of the procedure is most important to our audience, we sequence the procedure chronologically. For example, because the writer follows chronological order in the following procedure, the audience understands the progression of a second-opinion insurance action. The audience knows *when* specific people become involved as well as what they do:

When a program is established, it is not necessary to publish a list of the procedures on which a second opinion will be required. Instructions to the insured simply state that when he is told by his doctor that surgery of a nonemergency nature is required, he should promptly contact the plan administrator in person or by phone. He will be asked to furnish the administrator with the name of the attending physician, his own version of the medical problem involved, the date surgery is scheduled, his address, phone number and a choice of two preferred appointment dates and most convenient time.

On the basis of this information, the plan administrator decides whether a second opinion is necessary. If he feels that it is, he consults his directory of specialists, picks the appropriate specialist, and makes an appointment for the examination. He then notifies the claimant of the appointment time and sends him a presurgical screening form. The presurgical screening form must be completed promptly by the attending physician and the insured. The insured is required to authorize release of necessary information to the specialist. The attending physician is asked to indicate the surgical procedure, diagnosis, pertinent history, X rays and laboratory findings, where the procedure will be performed, expected date of surgery, date of hospital admission, expected length of hospital stay, and total surgical fee, including after care.

This form is submitted to the specialist who will make the second examination. He will consider not only the need for the surgery, but also the qualifications of the doctor who plans to do it. If hospitalization is planned, he will also determine whether such hospitalization and expected duration are necessary. He must have access to the X rays and lab work done by the attending physician. Prior approval of the plan administrator is needed before additional tests may be made.

Notice how many people must coordinate their efforts to carry out the procedure; the person who is insured, the doctor, the plan administrator, and the specialist all perform their specific duties during the overall procedure. The writer asumes that the plan administrator, the doctor, and the specialist are well trained in their fields; he shows them how to coordinate their efforts. In a chronologically ordered procedure, the writer not only recreates the procedure as it happens but also conveys to the audience that certain steps *must* happen before others.

Most procedures are chronologically ordered, but if in a procedure, the tasks or categories of tasks are more important than the exact order in which they are carried out, or if the order would vary from case to case, we could sequence the procedure developmentally.

# Selection in Procedures

## A General Procedure

As with all narratives, in procedure we select the amount and type of detail to include according to audience and purpose. If our audience knows a great deal about the subject, we might give just the general outline of the procedure and expect the audience to fill in the details. We can start such a procedure with core material rather than a thorough introduction. For example, in the following procedure, the writer includes two methods of assessing cutting force and path error in tools. The writer assumes that the audience can fill in specific details when actually following the procedure:

> . . . Two experimental approaches are possible; first, one can measure the force components with a tool dynamometer for a series of depths of cut. If in the manufacturing of a product, the cutting speed or other machining parameters change, tests are required also to include those changes. Feed rate and depth of cut are somewhat analogous but can be varied independently if feed rate changes are planned in the actual part manufacture. Using the resultant of the force components and the depth of cut, an effective spring rate for the tool-workpiece combination may be calculated. A record of these "cutting spring rates" can be kept since the primary variables are associated with the workpiece and tool materials when other cutting parameters are the same. The second approach is to replace the dynamometer with either a disc of known thickness or a cylinder of known dimensions. Then, using the theoretical deflection prediction or a measured deflection, the force can be calculated from measurement of the cutting depth error. The error is primarily a record of the deflections produced by those force components. Thus very simple geometries, whose theoretical deflections are well known and easily calculated are used to provide information useful in more complex parts. . . .

Obviously the audience of this procedure consists of experts in the field who can choose between and easily follow the two methods.

## A Specific Procedure

On the other hand, the writer of the next procedure includes much more detail. This writer clarifies and justifies the procedure so that the audience outside the company will cooperate. First he presents the overall procedure chronologically, and then he gives exact details about the audience's particular role—that of policyholder. The writer also explains the procedure from the beginning to the end of the audience's involvement. While the members of the audience need to understand the "core" of the procedure, they also need specifics about their own roles:

> We have found that there is some confusion regarding the handling of overpayment reimbursement checks and returned checks. It might be helpful to review the company's procedure when we receive one of these checks in our Accounting Section.
>
> When a reimbursement check comes in, several things are checked. First, they check to be certain that it has come through you, the policyholder. Second, they verify that the payee is the company. If it is not, if the payee is the policyholder, then they make sure that it has been endorsed by the policyholder. Also, they examine the validity of the check, that is, see that any expiration date on it has not passed. Finally, they check to be sure there is enough information to complete an adjustment brief. That information includes branch number, coverage code, draft number, date of draft, liability incurred date, first or other payment, and regular or maternity coverage.

If any of this information is missing, or is incorrect, then either the check must be returned or a request for information must be sent to you. Much time is lost in crediting the appropriate account or adjusting the claimant's payment when this occurs.

You can help expedite credit adjustments by carefully looking over any checks that are overpayment reimbursement checks. If the payee is the policyholder or is jointly the policyholder and the company, the check *must* be endorsed by the policyholder. It cannot be cleared through our Treasury Department or any bank until it is endorsed. If the check is invalid, if a 90-day expiration date has passed, *do not* send it in unless the maker has extended the expiration date and stamped the face of the check to show this extension, or obtain a valid check and send it in.

When writing procedures, we select and sequence details according to our audience and purpose. The audience of the procedure above needed enough detail about the procedure to complete its role and to understand the importance of following exactly company procedure. Audiences with more expertise may need less detail about a familiar procedure and more detail about a new one. Audiences outside a company may need less detail about procedures in which they are not involved directly and more detail about those in which they participate. We might include less detail in reviewing a familiar procedure and more detail in justifying a new one.

# Indicating Alternative Approaches in Procedures

Often we present options within a procedure. Since procedures describe generally how an operation is handled, we assume that the members of the audience know their own specific tasks. We trust them to judge how individual cases might fit into the procedure. However, we still suggest or clarify options or exceptions in a procedure, especially a new or complicated one.

For example, in the following communication, the writer explains a new procedure for handling maternity insurance claims based on a new federal law. Some exceptions must be followed, while other options are just suggested. With the latter, the writer explains the possible effects of choosing each option, but leaves the final decision up to the audience. If indeed the members of the audience make the decision, the writer prefaces these options with "we think," "it would seem," "you may find it easier," and so on. If these exceptions or alternatives *must* be handled in a certain way, the writer prefaces these statements with "will therefore be made," "will be waived," and so on. In a procedure, we must make clear the difference between options and requirements. Study carefully how these differences are expressed in the procedure:

Our understanding of the law is that any extension of benefits provision must apply equally to all illnesses. Consequently, the special pregnancy extension will be eliminated after the effective date.

Total disability at the time of termination of coverage and its continuation until termination of pregnancy and beyond (from that pregnancy) must be established *to the same extent as for any other illness.* To apply this literally could, however, result in an employee or covered dependent, already pregnant on the effective date, being left without coverage for expense incurred after the effective date, and ending with termination of a pregnancy on, before, or shortly after 7 months from the effective date.

Two exceptions will therefore be made. First, the requirement that total disability be present at the time of insurance termination will be waived and the current special extension for pregnancy will be retained on an administrative basis. The benefit payable will be at least as great as it would have been in the absence of the amendment.

Second, if the current maternity benefits under a plan would provide a *greater* benefit than if paid on a "regular" basis, the greater benefit can be allowed.

The rationale for both exceptions is simply that a person becoming pregnant prior to the effective date had a contractual expectation that certain pregnancy expenses would be covered.

For someone covered both under the old maternity benefit by administrative extension and under the new regular benefit, two sets of calculations may be necessary, whether bills are submitted on a periodic basis or at one time after pregnancy has terminated. In the latter instance, the calculations can be done in one stage. Where bills are submitted periodically over the course of the pregnancy, it would seem preferable to calculate initially on the basis of the new regular benefits and then, if necessary, make up any balance still due under the old maternity benefit after all known bills are in.

If this approach is taken, we think you should so advise the insured at the time, to avoid any apprehension on his or her part that benefits are being reduced or miscalculated.

As an alternative method of calculating in those few instances where the current maternity benefit is so large as to make it almost certain that regular benefits would fall well short of the mark, you may find it easier to calculate on the old maternity basis immediately.

. . .

While only an audience experienced in this procedure could understand and choose between options, the writer has distinguished clearly between options and requirements. In procedure we must make clear to our audience what we leave up to them and what they must do.

# Style in Procedures

Since procedures are a type of narrative, we use exact action verbs to indicate the steps in the procedure. We usually use the present tense of the verb, since the action is repeated, sometimes on a daily basis: "The plan administrator decides"; the clerks "examine the validity of the claim"; and so on.

In procedure, we usually suggest what "should" be done, how the audience cooperates to try to achieve a certain goal every time the procedure is followed. If we state what must be done in procedure or what is always done regardless of the case, we sometimes use the future tense of the verb: The section supervisor "will return to you the lower half to verify receipt of the check"; "special pregnancy extension will be eliminated"; and so on.

If the procedure itself is more important than who participates in it, or if it is a group effort and we are unable to identify precisely who participates, we might use the passive rather than the active voice: "Setback lines are also marked for the future"; "boundary lines must be determined"; and so on. However, before we choose the passive voice, we must make sure that our audience does not need to know who is involved.

Again we must make clear the differences between suggestion and demand. If we give the audience a choice of action, we should state each option and its effects completely.

Finally, since procedure is a special form of narrative, we must choose a point of view. Usually we use the third-person-objective point of view in which we can address the audience directly if necessary.

## Summary

A procedure is a type of technical narrative in which we tell many people how to coordinate tasks to achieve a common goal in an overall operation. Whether inside or outside the company, whether participants or observers, regardless of the direction of the communication, the audiences of procedures are usually familiar with the operation. We can sequence procedures chronologically or developmentally, and the amount and type of detail we include vary with audience and purpose. Also in procedures we need to clarify options and requirements. In procedures we standardize an operation, clarify or change an operation, or introduce a new operation.

## Specifications

Audiences of specifications are experts in their fields. Many audiences who use procedures and instructions do not read specifications. The audiences of specifications are contractors, designers, and engineers within an industry who may build highways or airplanes, bridges or automobiles, and so on. The direction of specifications then is generally lateral—from expert to expert in the same or different fields. Because of the "legal" nature of the specifications contract, we always use a technical vocabulary in specifications.

Since specifications are technical narratives, we must choose a sequence and the amount and type of detail to include in them.

## Sequence in Specifications

Since in specifications we tell how a project, a type of action, will be done or what the result of it must be, we apply the principles of sequence in narrative. We usually order the specific parts of a specification chronologically. For example, in the following specification, the writer prescribes the method to be used in the order the audience would follow naturally:

> Immediately after depositing concrete on the newly placed adhesive, the portland cement concrete filling shall be thoroughly consolidated until all voids are filled and free mortar appears on the surface. The concrete shall then be struck off to the required grade.

Ordering the detailed actions of specifications chronologically helps the audience follow and meet all the requirements.

However, the overall organization of specifications is usually developmental and moves from the general to the specific. We first define the project as a whole, as in the following example:

> The work to be done consists, in general, of constructing channel access ladders and grab bars as shown on the plans.
>
> Chain-link fence to be reconstructed and gates to be installed, concrete to be placed and such other items or details, not mentioned above, that are required by the plans, Standard Specifications, or these special provisions shall be performed, placed, constructed, or installed.

Then we give governmental and trade requirements. For example, we state employment practices and union requirements in the beginning of the specifications.

After we give general requirements, we organize the rest of the specification according to materials, such as "pavings, masonry, metals, timber, and roofing"; according to structures to be built, such as "fences, ramps, pillars, and surfaces"; or according to whatever topics are appropriate to the project or industry. Again, we sequence the specification topics from the general to the specific, such as "Section 10-1 Existing Highway Facilities to Be Reconstructed, 10.1A Reconstruct Chain-Link Fence, 10.1B Reconstruct Chain-Link Gate," and so on.

# Selection in Specifications

The amount and type of detail we include in a specification depend on whether we specify completely the exact way we want work done or mainly the end result of that work, for example, whether we tell the audience how to build the structure or what strength and dimensions we want the structure to have when finished.

## Prescriptive Specifications

Prescriptive specifications prescribe or state exactly how work must be done to reach the desired result. Prescriptive specifications are also called *closed* or *methods* specifications because in them we restrict by name the materials and methods to be used. An example of a prescriptive specification of materials follows:

> Cables shall be ¾-inch preformed, 6 × 19, wire strand core or independent wire rope core (IWRC), galvanized in accordance with the

requirements in Federal Specification RR-W-410c, right regular lay, manufactured of improved plow steel with a minimum breaking strength of 23 tons.

In the prescriptive specification above, the writer gives the audience no choice of materials. In the next example, the method is prescribed as well as the materials. The audience must use a certain type of epoxy in a prescribed way:

The drilled holes shall be clean and dry at the time of placing the bonding material and the steel dowel. The bonding material and the steel dowel shall completely fill the drill hole. The bonding material shall be epoxy adhesive of commercial quality conforming essentially to the requirements in Section 95-1, "General," and 95-2.03, "Epoxy Resin Adhesive for Bonding New Concrete to Old Concrete," of the Standard Specifications.

After bonding, dowels shall remain undisturbed until the epoxy adhesive has reached a strength sufficient to support the dowel. Dowels that are improperly bonded, as determined by the engineer, shall be removed. The holes shall be cleaned or new holes drilled and the dowels replaced and securely bonded to the concrete. Redrilling and replacing improperly bonded dowels shall be performed at the contractor's expense.

In prescriptive specifications, we prescribe exactly how we want the audience to perform the job and what exact materials they must use.

## Performance Specifications

Performance, also called *open* or *end result* specifications, prescribes the result of the project or often the performance capabilities of the structure to be built. This end result or performance must be measurable. For example, a performance specification might read: "Concrete in the highway foundations must have a comprehensive strength of 2800 pounds per square inch at 29 days of age." In performance specifications we generally leave methods up to the audience but state what these methods must achieve. For example, a performance specification follows:

The holes shall be cored by methods that will not shatter or damage the concrete adjacent to the holes.

Water and residue from core-drilling operations shall not be permitted to fall on areas occupied by lessees on the nonoperating right of way underneath the structures or public traffic, to flow across shoulders or lanes occupied by public traffic, or to flow into gutters or other drainage facilities.

The audience can choose what methods to use to prevent shattering or damage to the concrete and to prevent residue from falling on areas adjacent to the project.

Selection of detail in specifications then depends on whether we are writing prescriptive specifications, in which we include exact requirements of methods and material, or performance specifications, in which we dictate the end result of the project. Although in the past most specifications were prescriptive, now since so many new materials and methods can achieve the same end, we usually write performance or combine prescriptive and performance specifications.

# Expressing Alternatives in Specifications

As in instructions and procedures, in specifications we often must indicate alternative methods or materials to be used in a project. However, since most specifications are legally binding, these alternatives must be exact. The audience's options must be clear. For example, in the following specifications, the audience may use a pea gravel mix in constructing diaphragm bolsters, but only a certain mix can be used:

> Diaphragm bolsters shall consist of reinforced concrete blocks constructed at hinge diaphragms as shown on the plans and in conformance with the provisions in the Standard Specifications and these special provisions.
>
> Construction of bolsters shall conform to the requirements of Sections 51, "Concrete Structures," 52, "Reinforcement," and 90-10, "Minor Concrete," of the Standard Specifications. At the Contractor's option a pea gravel mix may be used which shall conform to the following:
>
> Each cubic yard of concrete shall contain a minimum of 658 pounds of cement and not more than 0.53 pounds of water per pound of cement.

The contractor has the option of using the mix or not; if the contractor does, the mix must be made of a certain material. In specifications, we must make clear what choices the audience does and does not have.

## Stating Responsibility

In addition, if the audience does have a choice in method or material, often we must state who is responsible, especially financially, should a possible method fail, such as in the following example:

### Alternative Methods of Construction

Whenever certain of the plans or specifications provide that more than one specified method of construction or more than one specified type of construction equipment may be used to perform portions of the work and leave the selection of the method of construction or the type of equipment to be used up to the Contractor, it is understood that the State does not guarantee that every such method of construction or type of equipment can be successfully used throughout all or any part of any project. It shall be the Contractor's responsibility to select and use the alternative or alternatives which will satisfactorily perform the work under the conditions encountered. In the event some of the alternatives are not feasible or it is necessary to use more than one of the alternatives on any project, full compensation for any additional cost involved shall be considered as included in the contract price paid for the item of work involved and no additional compensation will be allowed therefore.

We must make clear the degree of choice our audience has so that no questions arise during the project. In writing specifications, we must have foresight.

# Deriving Specifications

Up to this point, we have been concerned with how to write specifications. The specifications we have looked at were derived from standards set by previous jobs, by the engineers who investigated the project, and by trade associations and agencies. As engineers, designers, architects, or such, we might derive specifications from research and recommend or justify these specifications to an audience, whether or not we are involved in writing the final specifications. We might investigate a proposed project to recommend what should be included in the contract specifications. We might compare and contrast many alternatives before choosing the best one for the owner to require of the contractor. Since we would address the report of our investigations to a wider audience than that of the final specifications, our audience would be interested in our method of investigation as well as our analysis and recommendations. This audience might be composed of our supervisors, peers on the job, and the owners who might have contracted our firm to investigate the proposed project.

For example, in the following communication, the writer gives the method and results of an analysis of a project. The writer's recommendations are for the final specifications of the project:

### Spread Footings in Basements

Preliminary project concepts indicate that the proposed structure will have a full basement which will extend to approximate el 48. Satisfactory spread footings can be developed at el 46 in the very stiff Silty Clay stratum. The allowable net bearing pressure at this elevation is 4,000 psf.

Utilizing an allowable bearing pressure of 4,000 psf, we made settlement analyses of the stress distribution on the underlying stiff Clayey Silt encountered in boring 4 and on firmer soils found in the other borings. Based on a net load of 250 kips per foundation unit, we assumed an 8 ft square isolated footing for analytical purposes. In light of the recommended allowable bearing pressure our analyses indicate that approximately 1 in. of settlement can be anticipated in the southeast portion of the building, with less than ½ in. anticipated elsewhere. . . .

We recommend that the sheet piling be left in place permanently. This procedure would provide an additional barrier to keep water out of the basement and would prevent disturbance to the natural soil at adjacent footings that would occur if the piling is pulled.

Contract specifications should require that the Contractor submit sheeting and bracing design and sequence of operations for formal review prior to beginning any of the deep excavations within the building. It is essential that an approved design of the construction sequence be followed to avoid movement of the ground into the excavation and consequent loss of support for adjacent footings.

Although we may not write the final specifications, often we are involved in deriving specifications to suggest to an audience.

Our recommendations of contract specifications might also include the options the contractor might have in completing a project. Our recommendations help the specification writer prescribe one method or material, or alternative methods or materials.

## Style in Specifications

In writing specifications, we use simple, direct statements. Although our audience consists of specialists in the field, and we use technical terms to describe the project, we should write clearly so that no requirement is ambiguous. Using the same names for methods or materials each time they are mentioned also eliminates ambiguity. If we are recommending contract specifications, we need to justify our recommendations. We might use another rhetorical pattern, such as comparison-contrast, to support our recommendations. However, in the final contract specifications, we state, rather than explain, requirements.

While we write instructions in the imperative voice and we use "should" in procedures to express the goal of the operation, in specifications the words "shall" or "will" precede our verbs. The word "shall" indicates requirements, such as in the following:

> One sample of cable properly fitted with swaged fitting and right-hand thread stud at both ends, 3 feet in total length, shall be furnished the Engineer for testing for each 200 cable assemblies or fraction thereof to be furnished.

The contractor *must* supply test samples to the engineer or violate the contract.

The word "will" cannot be interchanged with "shall." In specifications, the word "will" is used only in statements about the future. In the following example, concrete bolsters *will* be measured and paid for because this action is part of job procedure:

> Concrete bolsters will be measured and paid for by the unit as diaphragm bolster. Bolsters to be paid for will be determined from actual count of the completed units in place.

Often "shall" indicates what the audience must do, "will" what the writer agrees to do.

We must use standard symbols and abbreviations as well as previously accepted terminology in specifications. For example, we should not represent seconds by "s" in one section and by "sec" in another. We often use terminology from similar, past specifications rather than creating new terms. Our requirements for dimensions and measurements must be exact, and we must indicate the conditions under which the measurements were taken. For example, if we were to state that a structure should weigh 3 tons, we would state whether we meant long tons (2240 lb each) or short tons (2000 lb each). We must state whether a substance should be tested or measured while wet or dry. In all cases we must prescribe how materials should be measured or tested.

Finally, the illustrations that accompany specifications are drawn by architects or experts in the industry. Again maps indicate work locations, and blueprints represent the proposed structure in great detail. The illustrations are just as exact and precise as the specifications must be.

## Summary

In specifications, a special type of technical narrative, we give particular requirements for proposed operations or jobs. Engineers or designers often derive specifications, and contractors are usually addressed in specifications. Specifications can be prescriptive and state exactly how work must be done, or they can be performance and indicate the end result of the project. Selection of type and amount of detail depends on whether the specifications are performance, prescriptive, or both. We usually sequence parts of specifications chronologically while we order the whole specification developmentally, by topics, from the general to the specific.

In specifications, we must state alternative methods or materials clearly and exactly. We use the word "shall" to indicate requirements while we use the word "will" to indicate normal procedure. Symbols and abbreviations have to be consistent in specifications, and we must state the conditions under which materials are measured and tested. Maps and blueprints help an audience visualize the details of the proposed project.

# Readings
## to Analyze and Discuss

## I

In order to form a marine company, orders must first be issued and passed through the chain of command. These orders specify the exact time, place, and facing position of the company formation. The time in the orders is the time when the company commander should receive the company report. Therefore, the forming of the company is begun 10 minutes prior to the time indicated in the orders.

To begin forming the company, the four platoons fall in by numerical order from right to left. Each platoon sergeant, with his back to the facing position, falls in his platoon. The platoon guide then initiates immediately the fall-in procedure which requires that the platoon be three paces in front of and facing the platoon sergeant and six paces from the platoon on the right; however, the first platoon guide centers the first platoon on its platoon sergeant. The remaining platoon sergeants center themselves on their respective platoons. Once his platoon is fallen in, the platoon sergeant receives the three squad leaders' reports in numerical order. When reporting, from his position of attention, each squad leader will render and maintain a salute until the platoon sergeant accepts his report and acknowledges his salute. After the third squad leader's report has been accepted, the platoon sergeant about faces and assumes the position of attention which indicates that he is prepared to report his platoon. The four platoon commanders then take their positions centered three paces in the rear of their respective platoons.

As soon as all the platoon commanders are in position, the company first sergeant centers himself between the first ranks of the second and third platoons, takes nine paces in the facing direction, about faces, and issues the command report. In numerical succession, the platoon sergeants report their respective platoons. When reporting, each platoon sergeant renders a salute, gives his platoon's report, and terminates his salute when it has been acknowledged by the company first sergeant. The company commander positions himself three paces directly behind the company first sergeant after the fourth platoon sergeant's salute has been acknowledged. The company first sergeant then about faces, renders a salute, gives the company report, and terminates his salute once the company commander acknowledges it.

Now that the company report has been received, the company commander issues his orders for the company to the company first sergeant. When the company commander is through issuing his orders, the company first sergeant salutes and terminates the salute after acknowledgement by the company commander. The company first sergeant then about faces and issues the orders to the company. The platoon sergeants then simultaneously render salutes which are acknowledged by one salute from the company first sergeant. The platoon sergeants issue the orders to and dismiss their respective platoons, thus terminating the company formation.

Company formation is a formal assembly of a military unit composed of four platoons and a staff of officers over these four platoons. Company formation is usually a 15-min to 20-min procedure. It is primarily a ceremonial procedure that symbolizes the unity and discipline of the military.

*Donald Thomas*

## *Questions for Discussion*

1. What is the overall goal of the procedure above? How many people are involved and what is their specific area of expertise?
2. Describe the sequence and amount and type of detail in the procedure.
3. Are any alternatives or options allowed in the procedure? Why or why not?
4. What would the audience and the people involved in the procedure be expected to know before reading the procedure?

## II

*Pumping Plant Equipment.* This work consists of modifying an existing electrical panelboard, replacing light fixtures, and furnishing and installing exhaust fans and a sump pump in conformance with the details shown on the plans and the provisions in Section 74, "Pumping Plant Equipment," of the Standard Specifications and these special provisions.

Light fixtures to be installed shall be weatherproof, die-cast, non-ferrous metal, chrome-finished, double-lamp fixtures with porcelain sockets, adjustable swivel arms, 150-watt PAR-38 flood lamps, mounting flange, gasket and cast ferrous metal outlet box.

Exhaust fans shall have open drip-proof motors and belt drive centrifugal blowers and shall be installed as shown on the plans. Motor and blower shall be assembled on a sub-base of welded construction by the blower manufacturer. The unit shall bear the Air Moving and Conditioning Association Certified Rating Seal. Blower shall deliver at least 1,300 CFM against a static head of ¾ inch and shall not load the motor to more than ⅓ horsepower. Blower shall have welded steel construction, baked enamel finish, 16-gage or heavier housing, and a rotating assembly which is statically and dynamically balanced. Motor shall be ⅓ horsepower, single-phase, 120-volt, 60 Hz., 1,750-pulley that can be adjusted to regulate the RPM of the fan to cover the ⅛ inch to ¾ inch static pressure range of the blower supplied. The sub-base shall be mounted with vibration insulators.

Sump pump shall be a heavy-duty vertical pump and motor complete with float type control and shall deliver, under test, at least 40 gallons per minute against a total head of 50 feet. Pump, as installed, shall not load the motor to more than 1½ horsepower.

Pumping plant equipment will be paid for on a lump sum basis.

The contract lump sum price paid for pumping plant equipment shall include full compensation for furnishing all labor, materials, tools, equipment, and incidentals, and for doing all the work involved in modifying, furnishing, and installing pumping plant electrical equipment, complete in place, as shown on the plans, and as specified in the Standard Specifications and these special provisions, and as directed by the Engineer.

## *Questions for Discussion*

1. What are the audience and purpose of the specifications above? Are the specifications performance or prescriptive specifications?

2. Are there any options or alternatives within the specifications? Are responsibilities assigned?

3. How are "shall" and "will" used in the specifications?

4. What types of graphic aids might support these specifications?

5. Describe the detail and the sequence in the specifications.

# Exercises and Assignments

1. **Problem:** A procedure recreates for an audience an action or operation in which many people are involved. Their specific jobs are coordinated to achieve a common goal. Often if we write a procedure in which we describe the many jobs of the people involved, we might also be asked to write instructions for each of these people. In these instructions, we would tell each person how to carry out a specific job. Reread Reading No. I at the end of this chapter. One of the main people involved in the procedure is the platoon sergeant.

Assignment:

Write instructions in which you give directions to the platoon sergeant in a marine company on how to carry out his specific role in forming the company. Use the proper format for instructions and add detail about the job if you wish.

2. **Problem:** Analyze a campus, local, or business organization in which you might be involved. Study the procedure used to reach one of the goals of the organization and the people involved in the procedure.

Assignment:

Write a procedure that recreates for an audience how the people within your organization work together to achieve a goal. Specify your audience and purpose for the procedure.

3. **Problem:** You have decided to have lunch at a local fast-food restaurant. You order a medium-rare cheeseburger and french fries. When your order arrives the cheeseburger is cold, burned around the edges, raw in the middle, and sitting in grease. You send it back to the kitchen. A few minutes later, the cook storms out of the kitchen and shouts, "If you don't like the way I cook a cheeseburger, you had better specify what you want." You decide to write prescriptive specifications for a medium-rare cheeseburger to give to the cook.

Assignment:

Write the prescriptive specifications. Be sure to use "shall" and "will" correctly.

4. **Problem:** You work in a large office building in a large city. Your firm, an engineering company, occupies the third floor along with an accounting firm. Both companies are small and share a reception area that faces the elevators on the third floor. Both firms have decided to remodel the reception area. Included in the reception area must be a receptionist's desk and chair, at least five chairs for clients waiting for appointments, a coffee table, and two floor lamps. Your supervisor has

asked you to analyze more specifically what will be needed in the remodeled area and derive specifications that might be included in the contract specifications.

## Assignment:

1. Analyze the needs and requirements of your firm and the accounting firm for the remodeled reception area. Add detail about the present reception area if you wish. Write an upward-directed communication to your supervisor in which you discuss your method of analysis and your recommendations.
2. Write the contract *performance* specifications for the reception area.
3. Write the contract *prescriptive* specifications for the reception area.

# PART FIVE

## Convincing an Audience

# 12

# Argument and Persuasion:
## Backing Ideas with Logical Reasoning and Provoking Action

*Distinguishing between a fact, opinion, and inference.*
*Using induction and deduction to support or derive an inference.*
*Using a syllogism to test the truth and validity of arguments.*
*Using and testing propositions and enthymemes in persuasion.*
*Using ethics, emotion, facts, and expert opinions in persuasion.*
*Appealing to the various audiences of persuasion.*
*Writing a proposal.*
*Reading and analyzing persuasive communications written by others.*

We are now going to learn how to convince an audience to accept and act upon our ideas. One method is argument or *logical reasoning*, a part of the overall technique of persuasion. When we attempt to convince anyone, we make assertions or statements of belief or judgment that invite controversy but that can be backed by argument or logical reasoning.

To argue at all, doubt and conflict must exist about the interpretation of our subject. For example, we can argue about whether the United States is entering a period of recession. We can analyze economic statistics, define recession, and then argue that our interpretation of the statistics leads us to conclude that the country is in recession. However, we cannot argue successfully about subjects that allow no interpretation. We cannot argue about the time of day, because we can easily verify the time by looking at a clock. We cannot argue successfully about whether lunch tasted good, because taste is subjective and cannot be the same for any two people. Thus we argue about subjects that must be interpreted and whose meanings are ambiguous or controversial. We assert a meaning that we back by logical reasoning; the two basic tools of argument or logical reasoning are *induction* and *deduction*. When we argue, we do not have enough evidence to prove absolutely our assertion, but we can use induction and deduction to support the logic or probability of our assertion.

Persuasion is usually based on a series of assertions backed by logical reasoning. The main assertion is called the *proposition* and states our essential interpretation of the subject and what we want our audience to do about it. However, in persuasion we want to eliminate the conflict and doubt that may have created the argument in the first place; we want our audience to agree with our proposition. While we use logical reasoning to support our proposition or to refute our opponents' assertions, we also use the techniques of persuasion—appeal to ethics and emotions, expert testimony or opinion, facts and statistics—to take our audience from conflict to agreement.

# Facts

A fact can be *verified* or proved by personal observation or by research in qualified sources. For example, we can verify the time of day by looking at the clock, by calling the time and weather number on the telephone, or by listening to the radio news. A fact is *universally* agreed upon or can be verified by anyone who wishes to repeat an observation, calculation, or experiment. Since facts can be verified, we might use them as evidence to support an argument, but a fact would not be the main subject of an argument. The following are facts that are universally accepted or that can be verified:

1. President John F. Kennedy was assassinated on November 22, 1963.

2. Mark Twain wrote *Adventures of Huckleberry Finn.*
3. Denver is the capital of Colorado.
4. The Colorado River flows southwest into the Gulf of California.
5. A triangle has three sides.
6. One pound equals 16 ounces.

## Stating Facts

When stating a fact, we must make certain that each *word* is precise and verifiable. Was there only one President John F. Kennedy? Are we certain that he was assassinated? Should we identify Mark Twain as the pseudonym of Samuel Clemens? Is there any other place called Colorado? Does the Colorado River flow in any other direction? Do equilateral, isosceles, scalene, right-angled, and obtuse triangles all have three sides? Should we say 7000 grams or 0.453 kilograms rather than 16 ounces? Because there was only one John F. Kennedy who was President in 1963 and by definition he was assassinated, because publishers still use Twain rather than Clemens, because no other state of Colorado exists, because the Colorado River flows southwest only, because any triangle by definition has three sides, and because our audience is more familiar with ounces than the equivalent grams or kilograms, our facts are well stated.

## Verifying Facts

We can verify facts by using our sense of sight, taste, feeling, hearing, or smell, or by using tools or instruments that aid our senses. For example, although we can observe that a certain table is rectangular, we need a scale to verify that the table top is 3 by 5 feet. We can verify facts from the past or facts based on observation we cannot make personally by researching qualified sources. For example, while we could not observe the end of the Civil War in the United States, we can read the eye-witness accounts of qualified observers or the analyses of well-known historians.

## Qualifying Facts

We often qualify facts by time or place. To say that "television is the most popular source of national news" is questionable unless we qualify our statement by a date and location: "In *1976* television was the most popular source of national news in *the United States.*" Other countries must rely more on the newspapers or radios for national news, and three decades before 1976 most people in the United States did not even own a television. We might further qualify this statement by citing our source: "According to the Smith poll, in 1976 television was the most popular source of national news in the United States." We should even define "popular"—"over 75 percent of the public depended mainly on television for national news." Facts then are limited to or qualified by time and place.

Finally, we can verify statements about the past and present but not about the future. We can only speculate about future events no matter on how much evidence or logic we base our speculations. We cannot verify that television will continue to be a popular source of national news until that "someday" comes. Because facts are statements about the past or present that can be verified by observation or research, we either prove or disprove them; we use them to support an argument, but we do not argue about them.

# Opinions

Our opinions are our personal emotions and attitudes, which are based on our experiences, education, and background. Since no two people, not even identical twins, have the same experiences, no two can have exactly the same opinions. Opinions then are dependent on the individual, and although people may have similar opinions, they are never identical. For example, two people may agree that John Brown is a good mayor, but they might interpret "good" in different ways. One may believe a good mayor is organized, the other economical. Even if the two people agree that "good" means "organized," one may recall an image of an "organized" father who kept his papers straight while the other may remember an "organized" teacher who accounted for every moment in the day. In opinions then, we express *subjective* or personal beliefs based on words that call to mind subjective images. The following statements are opinions.

1. *The New York Times* is the most honest newspaper in the world.
2. Going to an amusement park is boring.
3. Only Europeans appreciate fine food.
4. It is better to have loved and lost than never to have loved at all.
5. Nuclear reactors are unsafe.

What is honest to one person may be simply foolish or tactless to another. An amusement park may be boring to someone who works there but not to someone who has never visited one. What is meant by "fine" food—gourmet, fat free, rich? A love affair may be so devastating to one person that he or she would have preferred to avoid it. Opinions then are personal attitudes expressed by words that convey subjective meaning. Since our audience can usually identify an opinion as such, we often preface an opinion with "I think" or "I feel." These words make clear to our audience that we acknowledge our statement as opinion, not as fact or inference.

## Expert Opinions

Opinions can be expert opinions and carry weight in an argument. If Linus Pauling, Nobel Prizewinner, says that nuclear reactors are unsafe, we would respect his opinion. While an opinion is seldom the main assertion of an argument, we can back an assertion by expert or reliable opinions from leaders in a pertinent field.

Opinions can become facts if eventually verified. If in the next 25 years, we experience more and more nuclear "accidents," we might conclude logically that nuclear reactors are unsafe. Although we might verify an opinion so absolutely that it becomes a fact, more likely we will discover only enough supporting evidence to relabel our opinion an "inference."

# Inferences

By definition an inference is a logical conclusion. Assertions, the main issues of argument, are one kind of inference; scientific theories and hypotheses are also inferences. An inference is backed by sound logic, based on some facts, but not completely verifiable. Although it may be proved true or false eventually, when we state an inference, it is subject to conflict and doubt. An inference, a statement of probability, often is not qualified by time or place. An inference then falls in between fact and opinion; it is backed by too much evidence to be an opinion but by too little to be a fact.

The following statements are inferences:

1. Increasing natural background radiation by 10 percent would mean thousands of more cancer cases each year.
2. In the next 200 years, the United States will run out of oil.
3. If President Kennedy had canceled his November 1963 trip to Dallas, he would have lived to complete his term of office.
4. If you drop your watch in water, it will stop ticking.
5. Dinosaurs became extinct because of lack of food.

If natural background radiation today causes cancer, increasing this radiation probably would cause more cancer. However, scientists may find a means of preventing cancer or of screening our background radiation. The original inference is logically sound and probably true, but we cannot verify it completely. In the same way, we have evidence that we are running out of oil, but we can only speculate about the future. President Kennedy probably would have finished his term, but we cannot be sure that he might not have been assassinated in another city, or have resigned before completing his term, or that some other situation might have developed. Most watches stop ticking in water, but some do not. Finally, we often pose theories or inferences about the past. We have some evidence that dinosaurs ran out of food but no eye-witnesses to the event. We can infer but not verify.

Inferences then are logical conclusions backed by probability, evidence, and reason. Inferences posed as assertions are the issues in argument. But before we examine what part inferences play in argument and persuasion, we need to see how they are derived from logic.

# Inferences Derived from Logic

## Induction

Induction is a logical thinking process in which we move from some events, things, or situations of a certain class to an inference about all events, things, or situations of that class. We assume in induction that what we find true of the parts of the class is true of the whole class. This assumption or inference is the result of an "inductive leap," which while logical and probable cannot be verified. For example, if we sample a number of hard green apples and find that every hard green apple we bite into is sour, we can infer that all hard green apples are sour. The pattern of our induction would be as follows:

Number 1 hard green apple is sour.
Number 2 hard green apple is sour.
Number 3 hard green apple is sour.
. . . (up to a representative sample)
Therefore, all hard green apples are sour.

In induction we move from verifiable evidence or facts to an inference called a generalization. Because we cannot test every hard green apple in existence, our generalization, while logical and probable, is not completely verifiable.

## Induction as Example

We seldom cite all of our evidence or examples for induction. We may point to our most significant or common examples, just as we may state our generalization before or after our selected examples. For example, instead of presenting our whole pattern of induction for our assertion about hard green apples, we may say:

> All hard green apples are sour. I know that they are because I tested 1000 hard green apples from various locations. For example, over 50 hard green apples from a small orchard in Washington all proved to be sour, while in the famous Jones orchard of upstate New York, all 550 hard green apples tested were sour. Therefore, I conclude that all hard green apples are sour.

Although we infer that all hard green apples are sour after testing a representative sample, we may offer only the most significant examples in our discussion. If we present our entire sample, we might do so in a table or appendix. By using only representative examples, we emphasize our generalization.

## Uses of Induction

We also use induction to identify cause. Our evidence becomes the effects we can observe, and the inference or generalization becomes the cause. For example, if every time we eat a hard green apple, we get a stomachache, we might conclude that green apples caused stomachaches. Our conclusion becomes even more probable if our friends also get stomachaches every time they eat hard green apples.

Analogy can also be a part of the inductive process. In analogy in induction we infer that because two or more items share several similar characteristics, they probably share other essential characteristics. For example, we observe that both human infants and young monkeys depend on their mothers for physical protection, food, warmth, and a great degree of affection, and that both touch and hug their mothers as often as possible. If we then observe that when young monkeys are taken at birth from their mothers and are raised without affection, they become despondent and die, we can infer that under the same conditions, a human infant may die.

# Deduction

In deduction, we draw a conclusion or inference from a general reason and a particular fact. If these premises are verifiable and presented according to certain rules of logic, then our conclusion must be accepted by our audience. Generally, our conclusion is more specific than at least one of our premises, and that premise may be a generalization derived from induction.

For example, if in induction we sampled many hard green apples and found them to be sour, we infer that all hard green apples are sour. So if we find another hard green apple, then without even tasting it, we can infer by deduction that it too will be sour. Our reasons for inferring that it will be sour are that (1) all hard green apples are sour and that (2) we can see that this apple is hard and green. The pattern of our whole thinking process follows:

|  |  |
|---|---|
|  | **Induction** |
| Number 1 hard green apple is sour. | Evidence or examples |
| Number 2 hard green apple is sour. |  |
| Number 3 hard green apple is sour. |  |
| . . . (up to a representative sample) |  |
| Therefore, all hard green apples are sour. | Generalization becomes a reason or premise in |
|  | **Deduction** |
| All hard green apples are sour. | First premise |
| This apple is hard and green. | Second premise |
| Therefore, this apple is sour. | Inference based on deduction. |

The first premise in deduction is usually a generalization, the second premise relates a specific case to the first premise, and the conclusion makes an inference about the specific case.

## Syllogisms

The pattern of deduction is called a syllogism. The first premise, usually the generalization, is called the major premise and asserts a relationship between two classes, for example, hard green apples and sour things. The second premise, the minor premise, relates a particular thing, this apple, to one of the classes mentioned in the major premise, hard green apples. The conclusion, the inference derived from deduction, asserts a relationship between the particular thing, this apple, and the other class mentioned in the major premise, sour things. By placing the conclusion and the reasons for it in a syllogism, we can see the assumptions behind our inferences. We can test or *refute* (disprove) the deductive argument because one of the premises—usually the minor one—is not true; in other words, it is not a verifiable fact. We can *refute* the inference because one of the premises, usually the major one, is not valid or is not logically correct, such as a hasty generalization, not based on a broad enough sample.

## Syllogism as Reason

When we use deduction in persuasion, we usually present our conclusion as our stand on an issue and one or two premises as the reasons for our conclusion. When we read a persuasive communication that depends on deduction, we need to reconstruct the syllogism to test its validity. When we use deduction in our own communications, we need to test our own syllogism to ensure its validity. For example, in the following paragraph taken from a letter from a private company to a government agency, the writer uses deduction to support his argument. We can reconstruct the syllogism to test its validity:

> The principle of confidentiality is essential to the practice of occupational medicine. The physician-patient (employee) relationship depends upon the trust which is a direct outcome of the patient's confidence that medical information will be held confidential. Since the proposed access rule allows the release of identifiable confidential medical records, it can only serve to undermine this trust and tend to reduce the information available to the physician. Thus, the OSHA proposal will in fact work to the detriment of good occupational health practice.

The reasons for the conclusion appear at the beginning of the paragraph:

Reason:        The principle of confidentiality is essential in occupational medicine.

Reason:        The proposed Occupational Safety and Health Administration (OSHA) rule that allows the release of confidential records undermines this confidentiality.

Conclusion:           Therefore, the OSHA proposal will work to the detriment of the occupational health practice.

By placing reasons and the inference drawn from these reasons into a syllogism form, we can test or refute the syllogism; we can see clearly the assumptions behind the assertion that the OSHA proposal is detrimental. To test the syllogism, we can ask what evidence led the writer to conclude that occupational medicine depends on confidentiality. What observations led him to draw this generalization? What inductive pattern? Also we can test the truth of the statement that the OSHA rule allows release of confidential records. Is this a verifiable fact? Therefore, reconstructing the deductive and inductive patterns in an argument allows us to test or refute the truth and validity of the argument.

# Enthymemes

Not only do we have to reconstruct a syllogism when we test or refute deduction in a persuasive communication, but often we also have to supply the missing parts. Writers often imply rather than state all their reasons or let their audiences draw their own conclusions. This form of syllogism, with one premise or the conclusion implied but not stated, is called an *enthymeme.* The missing part may be a commonly accepted premise, or it may be the most vulnerable part of the syllogism. An enthymeme usually includes two parts of the syllogism in one sentence or two closely related sentences, such as follows:

If vitamin C cured John's cold, it will cure mine.

This ethymeme expresses one premise and one conclusion. Changing the verbs in these clauses to the verb "to be" and separating them, we have the following:

Premise:         Vitamin C is a medicine that cured John's cold.
Conclusion:    Therefore, vitamin C is a medicine that will cure my cold.

Since the one premise we have contains the subject of our conclusion, "vitamin C," or the minor term, we are missing the major premise. The entire syllogism follows:

Major premise:    All medicines that cure John's cold will cure mine.
Minor premise:    Vitamin C is a medicine that cured John's cold.
Conclusion:        Therefore, vitamin C is a medicine that will cure my cold.

When we place the enthymeme into the syllogism and supply the missing premise, we see that the major premise may be a hasty generalization drawn from too little evidence.

## Indicators of Enthymemes

Certain words indicate an enthymeme in persuasive communication. Premises are usually preceded by words such as:

| | |
|---|---|
| since | for the reason that |
| because | as shown by |
| assuming that | may be inferred from |
| whereas | may be deduced from |

Conclusions are preceded by words such as:

| | |
|---|---|
| thus | implies that |
| therefore | proves that |
| hence | leads us to the conclusion that |
| so | we may deduce that |
| consequently | shows that |

A persuasive communication often consists of a series of assertions that appear as enthymemes.

# Finding Enthymemes and Generalizations in Persuasion

Finding enthymemes and generalizations is essential in testing the logical cores of any persuasive communication. For example, the following two paragraphs appeared in a persuasive technical report and contain both enthymemes from deduction and generalizations from induction:

> Where an adversary relationship exists, all parties justifiably tend to maintain defensive postures since the risks are considerable. The owner risks large claims or adverse court settlements, major delays in completion of the system, increased total cost and public and higher agency criticism or loss of confidence. All of this may endanger completion of a major component of the system or the entire system. The engineer risks large adverse court settlements, higher and perhaps prohibitive cost of liability insurance, damage to his reputation and possible bankruptcy. The contractor risks major unanticipated costs, higher and perhaps prohibitive cost of bonding and possible bankruptcy. All parties may suffer from long, costly and debilitating litigation which has no public benefit but encourages further hostility and conflict.
>
> Since the real engineering of the tunnel occurs during construction operations, the engineer must provide a framework for modifying construction methods during the construction period with an equitable basis for sharing responsibility without excessively defensive postures and unfair gains or losses. Since much underground construction is time-dependent, such a process implies important decisions in the field. As an example, a silt deposit is judged by geotechnical engineers to have the capability of remaining stable at the tunnel roof for a number of hours. A team representing the owner-engineer-constructor can make the decision to work overtime on a calculated risk basis

to rapidly mine that portion of the tunnel without supports. If successful, the cost and time savings for the project may be appreciable and benefits may be shared by the owner and the contractor. If the effort is unsuccessful, the risk can similarly be shared. . . .

The first paragraph seems to be based mainly on induction:

The owner risks. . . .
The engineer risks. . . .
The contractor risks. . . .
Therefore, all parties take financial risks in the construction trade.

The inductive pattern is followed by a series of enthymemes which we can convert into syllogisms.

The first enthymeme relates to the generalization in the first paragraph: People in the construction trade are defensive and hostile because of the financial risks involved, or as a syllogism:

| | |
|---|---|
| Major premise: | All people who risk are defensive and hostile. |
| Minor premise: | People in the construction trade are people who risk. |
| Conclusion: | Therefore, people in the construction trade are defensive and hostile people. |

The next two enthymemes open the second paragraph: Since the real engineering takes place at the construction phase, shared responsibility must occur there; since underground construction is time dependent, important decisions must be made in the field. Converted to syllogisms the enthymemes appear as premises and conclusions:

| | |
|---|---|
| Major premise: | The real engineering phase is the time when shared responsibility must occur. |
| Minor premise: | The construction operation is the real engineering phase of constructing a tunnel. |
| Conclusion: | Therefore, the construction operation is the time when shared responsibility must occur. |
| Major premise: | If a job is time dependent, important decisions must be made in the field. |
| Minor premise: | Underground construction is a time dependent job. |
| Conclusion: | Therefore, in underground construction, important decisions must be made in the field. |

The examples that follow these two enthymemes support the two conclusions.

However, another enthymeme which we can draw from both paragraphs is the main enthymeme or the *proposition* of the persuasive report: Since sharing responsibility decreases hostility, the engineer can decrease hostility in the construction trade:

| | |
|---|---|
| Major premise: | Sharing responsibility is a thing that decreases hostility. |

Minor premise:   In the construction trade, the engineer is a person who can provide a means for sharing responsibility.

Conclusion:   Therefore, the engineer is a person who can decrease hostility in the construction phase.

The proposition states the main or most essential inference we make about our subject and often indicates the action we want our audience to take. The writer of the passage above implied that the engineer *should* therefore take steps to decrease hostility in the construction phase.

Once we have placed the examples and generalizations into an inductive pattern and the premises and conclusions into a deductive pattern, we can test the logical validity of any persuasive communication.

# Other Techniques of Persuasion

While persuasion depends on logical reasoning and while we often present our stand on an issue in a series of enthymemes, we use other important techniques to convince our audience to accept and act upon our ideas. We can also persuade our audience on the basis of ethics, emotion, facts, and expert opinions.

## Ethics

In persuasion we can appeal to the ethical standards of our audience or establish our own ethics or authority.

When appealing to the ethics of our audience, we establish a common goal. We convince our audience that the goal is morally right or generally beneficial. We make our appeal in the introduction or conclusion of our communication.

Much more common in technical communications is establishing the writer's own authority or ethics. Our audience will accept more readily what we propose if we first prove that we are experienced in the subject and are objective. If we are not writing out of self-interest, our audience will accept our recommendations and evaluations as being fair. For example, in a report to the Environmental Protection Agency, a private company first established its objectivity and good intentions before arguing for the use of a product:

> [Our company] is a major supplier of hydrocarbon propellants. We do not produce aerosol products; however, we do operate a technical service center that provides extensive technical assistance to our customers. In addition to helping our customers choose the most suitable propellant system for their product, our technical service personnel provide advice and formulation assistance to develop a finished product that is effective and safe for consumer use.

In a technical communication we can also establish our authority by reviewing our credentials.

## Emotion

While we seldom use emotion in technical communications, we should be aware of it as a persuasive device. The risks in using emotion in persuasion are similar to those of telling a joke; the audience responds well or not at all. Although emotion can be an effective technique in persuasion, audiences may have the opposite reaction we desire. They may so resent being manipulated emotionally that they ignore our logic.

## Facts and Statistics

While we argue about inferences rather than facts or opinions, we present what facts we do have to back our inferences. In technical communication, our facts, those statements we can verify completely, are often statistics.

Often the verification for our facts comes from our own experiments or research projects. In this case, our methods of verification must be objective, thorough, and professional. For example, in a professional paper, the 1979 Committee Chairman on Registration of Engineers of the American Society of Civil Engineers analyzed the results of a survey of engineers in Iowa. Before he could conclude that engineers who belonged to a professional society generally earned more money per year, he had to present his verification:

**Society Membership Versus Annual Salary**

| Do Belong to Society | Annual Salary | Do Not Belong to Society |
|---|---|---|
| 4.8% | < $20,000 | 14.0% |
| 40.4% | $20,000–$30,000 | 52.0% |
| 28.7% | $31,000–$40,000 | 24.0% |
| 26.1% | > $40,000 | 10.0% |
| 100.0% | | 100.0% |

Nearly two-thirds (66%) of all respondents who *do not* belong to a technical or professional society have salaries of $30,000 or less per year. Less than one-half (45.2%) of those respondents who *do* belong to a society have salaries in this range. By contrast, more than one-half (54.8%) of society members have salaries in excess of $30,000 per year; only slightly over one-third (34.0%) of those who are not society members have salaries in this range. . . . Whatever the reasons may be, it is a fact that civil engineering registrants in Iowa, who belong to one or more technical or professional societies, enjoy significantly higher salaries than those registrants who are not society members. . . .

The data were gathered from the results of a questionnaire and supported the writer's conclusions about the benefits of society membership.

A thorough presentation of facts that support our inferences and the verification for these facts greatly strengthen persuasion.

## Expert Opinion and Testimony

Although we cannot argue successfully about opinions, since they are based on subjective feelings, we can use expert opinion or testimonies from professionals in a relevant field to support an inference. We must be certain that our "expert" is an expert in the subject we are debating. We would be impressed with Linus Pauling's opinions on nuclear reactors but not on rock music.

For example, in an article on experimenting with animals, this writer quotes from a number of experts whose authority she establishes:

> "Some knowledge can be obtained at too high a price," writes British physiologist Dr. D. H. Smyth in his recent book *Alternatives to Animal Experiments.*
>
> "The lives and suffering of animals must surely count for something," says Jeremy J. Stone, director of the Washington-based Federation of American Scientists, which has devoted an entire newsletter to a discussion of the rights of animals.

While the statements of these individuals are still opinions, they can influence an audience who respects the experts' work in a field.

## Refutation

One essential task in persuasion is to refute the other side of an issue. We have seen how to refute an argument on the basis of logical validity. Also we have seen that verification establishes truth or falsity of facts. In persuasion we cannot write in a vacuum. We cannot present only our stand on an issue and ignore other possible alternatives. To convince our audience to accept our recommendations, we have to prove not only that our recommendations are feasible but that other choices are not. Regardless of whether we use truth, validity, or both to refute the other side, we must do so *before* our audience can. As most audiences read a persuasive communication, they naturally raise objections to our ideas. We must acknowledge these objections and refute them.

# The Pattern of Persuasion

Before we discuss the audiences and purposes of persuasion in technical communication, we need to study the traditional pattern of persuasion. Persuasion convinces an audience to be concerned about an issue, that one stand on the issue is better than others, and to act upon that stand.

In the beginning, we need to stress why our audience should pay attention to us. What benefits will they gain if they adopt our position? What specific needs must be met? Within the introduction, we also state our proposition. Again this proposition is our main assertion. In our proposition we identify the issue with which we are concerned and our stand on the issue. If possible, we should also state what specific action we would like our audience to take. Finally, in the introduction we may preview our proof; showing what points we will raise to support our stand helps our audience follow our reasoning.

The main body of our communication is the proof of our proposition. We can use logical reasoning (induction and deduction) as well as the other techniques of persuasion (facts, expert opinions, ethics, emotions, and refutation) to prove our point.

We end our communication with a summary of our main conclusions and a final appeal to our audience to act. The final appeal may depend on ethics, refer back to benefits, or be a formal recommendation of a concrete act.

This pattern of persuasion varies little from audience to audience. In convincing any audience, we always need to capture their attention, state our proposition, prove that we are right and others wrong, and urge our audience to accept our proposition. The field of advertising has represented this pattern by the abbreviation AIDA: attention, interest, desire (by proof), and action. The traditional pattern of persuasion is also used in technical communications.

# Audiences and Purposes of Persuasion in Technical Communication

Persuasion appears in almost every technical communication. Every time we explain a procedure, define a term, give instructions, describe a device, analyze a cause, and so on, we try to convince our audience that our ideas are well reasoned and our recommendations the best action to take. Persuasion then can be directed toward all audiences, and the techniques of logical reasoning and overall persuasion can be used to influence those audiences. While we have seen some examples of persuasion in technical communication already, a brief survey of the technical audiences of persuasion shows how widely used it is.

## Persuasion to Recommend Action to Supervisors

Most formal reports end in logical conclusions and recommendations. Whenever we investigate a problem, we analyze possible solutions before recommending one of them. Our entire report is usually persuasive, and we review our proof at the end of the report. In the next chapter, we study how to draw conclusions and recommend action to supervisors in a formal report.

## Persuasion to Recommend Techniques to Peers

When we conduct research in our specializations, we often report on new techniques to our peers in the field. As we studied in Chapter 1, we can suggest to but not command peers. In reports on our research, we often try to convince our peers that a new technique works well, that a discovery is accurate, or that a new device is better than the old one. We use the pattern and techniques of persuasion.

## Persuasion to Sell New Ideas to Employees

We may think that we need not persuade employees to follow our directions, since they must do so to keep their jobs. However, if we can persuade them that our methods are safer, faster, or easier, they will use them with more enthusiasm. We often sell ideas to our employees much as we sell a product, and we use the pattern and techniques of persuasion.

## Persuasion to Sell Products

Of course, we use persuasion to sell to customers the goods and services we produce. Facts, expert opinions, logical reasoning, and the other techniques of persuasion all help sell our products. Most sales material and advertisements capture audiences' attention, prove that our products benefit them, and explain the way and urge them to buy the products.

Persuasion is not only a "pattern," such as description or definition; it is also a "purpose," such as to inform. Most technical communications contain elements of persuasion.

# The Proposal

One special form of persuasion directed toward a supervisory audience is the proposal. The proposal asks for permission to undertake a certain project, within a certain length of time, for a certain sum of money. In a proposal we must convince our audience that we are the best people to handle the project. Our proposal must be logical in that we offer proof or evidence that we can undertake the project successfully. We must suggest a specific solution to the problem or plan for the project, an analysis of the problem or project, and a means of reaching that solution. Although there are many plans of organization for proposals, most contain a statement of the problem; a recommendation or proposed solution; the scope or approach to the solution; the personnel, funding, and time schedule involved; the experience and expertise of the personnel or company involved; and a final justification of handling the project. Usually proposals are written by employees to supervisors within a company or by company representatives to firms that might hire that company.

For example, the following proposal analyzes an internal problem, suggests the probable reasons for the problem, and proposes a study of one solution. The writer identifies who will authorize or give permission for the project, the subject of the project, the purpose, the scope or range, the suggested time limit, and the method or design of research and analysis:

MERCEDES MANUFACTURING CORPORATION

March 21, 1979

TO: Thomas P. Clarkson, Vice-President of Manufacturing
FROM: Martin Knox, Director of Machining Division
SUBJECT: Project Proposal for Investigation of Numerically Controlled Machining

Last year profits for the machining division were compatible with the 1977 profits, but the sales rose by more than $450,000. Why didn't profits increase proportionally? The answer lies in the manufacturing methods. Recently the orders have been in large quantities and the parts have been of a more complex nature. To keep up with an ever increasing production schedule, the machining division has installed 5 lathes and 6 milling machines and has employed 25 machinists to keep the machines busy for 2½ shifts a day. Because of the increased machines, the machine shop has reached its expansion limits; it is also important to note that the 25 machinists were extremely difficult to find.

The problem is that in the next 3 years, sales are expected to double. In its present mode of operation the Machining Division will be unable to increase production to a level that high. The prime objective of this proposal is to receive permission to study one possible solution: Numerically Controlled Machining. I feel that this new type of machining will enable the machine shop to increase its production without increasing its floor space, which will save the company money and time and increase profits as sales increase.

| | |
|---|---|
| 1. AUTHORIZATION: | Thomas P. Clarkson, Vice-President of Manufacturing; Jon J. Brand, Vice-President of Finance; and Anthony S. Salmon, Vice-President of Sales and Personnel |
| 2. IDENTIFICATION: | Numerical Control vs. Conventional Machining |
| 3. PURPOSE: | To gather data in order to evaluate the advantages and disadvantages of numerically controlled machining contrasted to conventional machining. |
| 4. SCOPE: | The main areas of the proposed study will be short-run orders, long-run orders, set-up, complexity, computer facilities, time studies, employee selection, quality, machine capabilities, and cost. |
| 5. LIMITATIONS: | Two-month period after acceptance of proposal. |
| 6. DESIGN: | a. Assumption: Numerically controlled machining will increase production with a smaller staff of unskilled employees contrasted to conventional machining practices. |

b. Tentative Conclusions: The proposed study will indicate when numerically controlled machining will be more advantageous contrasted to conventional machining.

c. Data Needed: The data needed will be primarily financial statistics but will also include data on the different kinds of machinery, employees, and size of orders.

1. Sources: Personal observations and interviews with companies that use numerically controlled machines, numerical and conventional machine salesmen and pamphlets, and information from industrial magazines and books.

2. Data Collection and Sequence: Library research, interviews, and observation.

3. Data Use: All information will be integrated into a formal report where conclusions will be drawn on the practicability of numerical machining versus conventional machining.

4. Report Format: Formal typewritten report with oral presentation.

5. Schedule: Progress report as requested, and final report due on or before May 15, 1979.

7. JUSTIFICATION: The information generated from the proposed study will provide a good foundation of the merits and disadvantages of numerical machining. It is hoped that this information will be instrumental in solving the problem of increased sales in a plant which is already operating at full production.

The proposal is one of the most common and important types of persuasive communications. In a proposal, although the format varies greatly from company to company, we ask for permission to tackle a certain problem in a certain way—we convince our audience to give us authorization to do so by logical reasoning and the other techniques of persuasion.

## Summary

Argument is part of the overall pattern of persuasion. We argue not about facts or opinions but about assertions, which belong to the broad category of inference, conclusions derived from logical reasoning. An inference can be based on induction, drawing a generalization from specific facts, or deduction, reaching a specific conclusion by applying a general reason and a specific fact and expressed in a syllogism.

In persuasion, we try to eliminate the conflict raised in argument and convince our audience to accept and act upon our assertions. We use the examples and generalizations of induction and the reasons and conclusions of deduction to support our assertions. Often deduction appears in the form of a modified syllogism or enthymeme. We recreate the syllogism to test its validity and to verify its premises. The main assertion in persuasion, a proposition, states our stand on the issue and the action we recommend and is usually stated in the form of an enthymeme.

The other techniques of persuasion are ethical appeals and establishment of authority, emotion, facts or statistics and their verification, expert opinion and testimony, and refutation. The pattern of persuasion—getting the audience's attention, convincing them to care about the issue, proving our stand on the issue, and calling for specific action on the issue—is appropriate in technical communication.

In technical communication, often we use persuasion to recommend action to supervisors, to recommend techniques to peers, to convince employees to follow procedures, and to sell products to customers. The proposal convinces an audience to authorize a project or let us investigate a solution to a problem.

# Readings
## to Analyze and Discuss

# I

In a letter of April 26, 1979, we submitted comments on the Environmental Protection Agency [EPA] "Clarification of Final Rule" issued on November 27, 1978, regarding chlorofluorocarbons. This letter explains further our interest in nonpropellant use of CDC-183.

[Our company] is a major supplier of hydrocarbon propellants. We do not produce aerosol products; however, we do operate a technical service center that provides extensive technical assistance to our customers. In addition to helping our customers choose the most suitable propellant system for their product, our technical service personnel provide advice and formulation assistance to develop a finished product that is effective and safe for consumer use.

Our safety consideration is that of flammability. Although we do not manufacture CDC-183, we know that it does provide flame retardant properties to certain formulations. Since the Federal Register notice of March 17, 1978, specifically cited CDC-183 as an example of a chlorofluorocarbon that would not be subject to regulation, our technical service personnel continued to study the use of CDC-183 as a flame retardant in many types of products.

CDC-183 offers advantages over other materials commonly used as flame retardants in aerosol products. Since it has no odor, CDC-183 can be used in perfumes, deodorants, and antiperspirants as well as insecticides and other types of formulations. In addition, CDC-183 does not attack or etch plastics and thus can be formulated into products such as those used in fumigating international aircraft as required by the WHO.

Use levels of flame retardant chemicals in aerosol products vary with the flammability characteristics of the active ingredients, the solvent, and the propellants. A comparison of the effectiveness (as measured by the decrease in flame extension) of CDC-183 with methylene chloride indicates that the CDC-183 provides greater flame retardancy for the same use level.

CDC-183 is a liquid at room temperature, has a low vapor pressure, and can be easily added to an aerosol product along with the active ingredients and solvent portion of the formulation. This mixture of active ingredients, solvent, and CDC-183 is placed in the container by the filler and then the appropriate type and amount of hydrocarbon propellant is charged to the container. Thus, the CDC-183 enters the product with the product concentrate, and the propellant is added in a separate final step.

In light of this additional information and our earlier comments, we urge the Agency to consider this proposed non-propellant use of CDC-183.

## Questions for Discussion

1. What are the audience and the purpose of the letter?
2. What is the proposition of the letter? What is the main assertion about CDC-183, and what action does the writer want the EPA to take?

3. What facts, opinions, and inferences support this proposition?

4. What patterns of induction and deduction are used to present the evidence?

5. What other rhetorical patterns support the argument?

# II

April 30, 198__

Mr. Jack Cransford
Director
Roget Corporation
27 Main Avenue
Denver, Colorado 80210

Dear Jack:

Thank you for arranging our meeting with Larry Smith to discuss his current recruiting requirement for an Associate Corporate Comptroller-Financial Planning. We welcome the opportunity to participate with you and Larry in this recruiting effort.

The purpose of this letter is to summarize our understanding of this position, develop our participation in the engagement, and outline time requirements and related professional fees.

## The Position

Reporting to the Comptroller, Mr. Smith, the Associate Comptroller will be responsible for the preparation and review of the Corporation's annual budget, the supervision of the monthly forecast and analysis of the Division's operations, the presentation of operating results, the supervision and review of special project information utilizing the financial analysis staff and selecting, developing, and training new MBA analysts for eventual placement in the Divisions. It is further understood that the Associate Comptroller will work in close conjunction with his counterparts in the Accounting and Management and Planning areas during their times of peak workload.

Acceptable candidates will possess eight to twelve years of financial management experience which should encompass a period of time with a major public accounting firm, a position of similar responsibility in financial planning and analysis with an industrial manufacturer or a Division Controllership responsibility with a smaller organization. Financial systems consulting could be substituted in this experience quotient. It is most desirable that candidates possess a CPA and an MBA degree.

Hiring range has been established in the $55,000 to $60,000 area to include a competitive corporate benefits program. We expect the successful candidate to possess the interest to progress into other areas of financial management and eventually qualify as backstop to the Comptroller.

Under my direction, RR Company will conduct this search. We will contact known sources, interview selected individuals, and check professional references of qualified candidates. As acceptable individuals are identified, they will be presented to you for evaluation.

### Time Requirements and Related Professional Fees

Should you select RR Company to perform this search, we will gear our efforts to complete it in a 75- to 90-day period. Our professional fees for services rendered equal 30% of the agreed-upon first-year's total guaranteed compensation paid to the hired candidate. When approved, we will direct to you our initial invoice in the amount of $4,000 with a second and third invoice of equal amount to be forwarded to you after thirty and sixty days respectively. The balance, the difference between $12,000 and 30% of the first-year's total compensation, will be billed to you at the completion of the engagement.

Additionally, we bill for out-of-pocket expenses incurred during our search, including telephone, meals, and approved travel expenses for candidates and RR Company associates. It is agreed that you are at liberty to terminate our services at any time during the search upon payment of the first and second installments due plus the out-of-pocket expenses accrued to that point.

We thank you for arranging our meeting with Larry and requesting our proposal on this most important search. Please call me with any questions or comments on this proposed work schedule. We look forward to working with you through to the successful conclusion of this engagement.

## Questions for Discussion

1. What are the audience and purpose of this proposal? What patterns of induction or deduction or techniques of persuasion do you find here?

2. How does this proposal format differ from the one in the chapter? Can you justify these differences?

3. Why does the writer introduce the proposal with a summary of the information gathered at the last meeting?

4. If you were Jack Cransford or Larry Smith, would you hire RR Company for this job? Why or why not?

# *Exercises and Assignments*

**1. Problem:** Study Figure 12-1 and identify the following statements as fact, opinion, or inference:

**a.** The scene depicted in the photograph takes place on a national holiday.
**b.** The woman facing the camera is carrying a purse.
**c.** The truck and crane are parked on the sidewalk.
**d.** The scene depicted takes place in a small town.
**e.** One of the men wearing a hat is shorter than his friend.
**f.** The man on the truck is replacing a broken fire escape.
**g.** When this photograph was taken, it was summer.
**h.** The photograph is rectangular.
**i.** Parking trucks on city sidewalks is dangerous.
**j.** The two women in the foreground wear striped dresses.
**k.** The man sitting on the truck is waiting for something.
**l.** There are several stores behind the truck.
**m.** Two men watching the truck and crane wear hats.
**n.** The two men with hats are curious about the truck and crane.
**o.** The two men with hats are retired.
**p.** Our senior citizens don't have enough interests to fill their time productively.

*Figure 12-1.* Photograph by Faye Serio.

**2. Problem:**   Evaluate the following paragraphs as examples of inductive reasoning. Are the generalizations supported by the evidence? Are there any fallacies in the reasoning?

**a.**  Last year my roommate got a new portable electric typewriter for Christmas. In March the carriage stuck and he had to get it fixed. Then in May he had to replace the margin release. My father says that since his office switched from manual to electric typewriters, their repair bills have increased by 25 percent. I've had a manual typewriter for 5 years and never had a problem with it. Electric typewriters are less reliable than manual ones.

**b.**  In a recent survey of cancer victims, 69 percent of those interviewed stated that they had experienced long periods of depression and anxiety at least a year prior to being diagnosed. A secure and happy life helps prevent cancer.

**c.**  Most landlords in this town prefer to rent to students rather than to families. The student housing association recently responded to all the ads in the local paper for apartments and houses with two or more bedrooms. For half the calls made, the associate director called first and identified herself as a student with roommates. Twenty out of 23 landlords agreed to interview her. Then the housing director called and identified himself as the head of a family. Only 11 landlords wanted to interview him. Then the director and associate director reversed the order of their calls. Only 13 out of the second set of 23 landlords would see the associate director as head of family while 21 would see the director as student.

**d.**  Allowing a nation to continue in a state of inflation is like spoiling a child. Like the child who gets everything, the nation that spends without limit never learns the value of money. Rather than accepting that some things are not available, both the child and the nation seek other ways to obtain that thing without regard for consequence. Neither the child nor the nation build moral character to endure hardship or learn the rewards of patience. So like the parent who must set limits for the child, the government must set limits on our spending.

**3. Problem:**   Analyze the following two generalizations and the evidence listed below each. Which pieces of evidence support the generalization? What other evidence is needed? Should the generalization be modified?

**a.**  During the next decade, the number of college students studying the pure sciences will decrease.
   1. Most industries now require scientists to have at least a master's degree.
   2. America is more suspicious of science and technology now than ever before.
   3. Engineers earn more money in their first year on the job than scientists do in their first 5 years.
   4. It is very difficult to get into medical school now.

5. Enrollment in the School of Arts and Sciences at the 10 largest universities in this country decreased by 8 percent this year.
6. Last year 10 percent fewer students studied biology at the five top-rated liberal arts schools.
**b.** People who take drivers' education courses are safer drivers than people who do not.
1. Three major insurance companies give a 3 percent discount on premiums to drivers who have taken the course.
2. Seventeen states require that all drivers take the course before receiving their licenses.
3. In this state, people who have their licenses revoked must retake the course.
4. Of all those people involved in a major accident in this state this May, 45 percent had never taken the course.
5. The course teaches defensive driving skills and sharpens reaction time.

**4. Problem:**  Place each of the following statements into syllogism form and test their validity.

**a.** All democratic forms of government tolerate political discussions. America tolerates political discussions because it is a democracy.
**b.** We must either win this war or die! We will win the war and live!
**c.** The college must build a new parking area if it builds a new dormitory. The plans for the new dormitory were abandoned this year, so we won't need any additional parking space.
**d.** Some professors really stay students for their entire careers. Because some students tutor other students, some students are really professors.
**e.** You're either my friend or my enemy. Since you are not my friend, I have put you on my list of enemies.
**f.** Some students will never finish college. Because some people who don't graduate from college get excellent jobs, some students will have successful careers.
**g.** Criminals make good detectives because criminals are suspicious people and good detectives must be suspicious.

**5. Problem:**  Place the following enthymemes into the syllogism form, supply the missing major or minor premise or conclusion, and test the validity of the syllogism.

**a.** Since careless drivers die young, John will not reach the age of 30.
**b.** All politicians are ambitious, and John is a politician.
**c.** All politicians are either honest or dishonest, and John is not dishonest.
**d.** Since some politicians live in Washington, John must live there.
**e.** Because Harold is an Independent, he cannot vote in the primary election.

**f.** Most redheads are passionate; therefore, you must be a very passionate woman.

**g.** The temperature has dropped to 20 below zero. If it's that cold, my car won't start.

**h.** If we want to cure cancer, we must find the cause, and so far, we haven't found it.

**i.** From the rising cost of living calculated this month, we can deduce that the unemployment rate will continue to be high this year.

**j.** All watches need to be cleaned every two years, and my watch really needs a good cleaning.

**6. Problem:** Your company, which manufactures snowmobiles, had decided to institute a 4-day, 9-hour-a-day work week for the months of October to March. The rest of the year the employees work a 5-day, 7-hour-a-day week. The company is located in Northern New York near the Canadian border and wants to cut down on fuel costs in the winter but keep up with the seasonal demands for the product. Your supervisor has asked you to draw up a list of advantages and disadvantages to this plan. You may add detail about the company, the product, and the location.

**Assignment:**

**1.** Draw up a list of advantages and disadvantages to the plan.

**2.** Write a brief report either supporting or refuting the plan. Recommend definite action to your supervisor.

**3.** Go back and annotate your communication in the margins of your paper. List the patterns of induction and deduction as well as the other rhetorical patterns you use. Note these patterns in the margin next to where they appear in your paper.

**4.** Analyze your annotations and revise your paper to improve your use of the patterns of argument and persuasion as well as of any other rhetorical patterns.

**7. Problem:** Choose an issue with which you are familiar and take a stand on that issue. It may be a campus problem, a political or economic topic, or a subject in your technical or business field. Gather evidence to support your proposition.

**Assignment:**

**1.** Write a proposal to your hypothetical audience to convince them that you are the best person to investigate the issue.

**2.** Write a persuasive communication in which you state and defend your proposition. Decide upon a specific audience and purpose. Use the appropriate techniques of argument and persuasion, and recommend concrete action.

# PART SIX

## Choosing or Creating a Format for a Communication

# 13

# Using Conventional Formats and Rhetorical Patterns to Organize a Letter, Memo, or Report

*Writing a personal and positive business letter that conveys a "you" attitude.*

*Combining the conventional format with a created format in a letter.*

*Writing a concise memo.*

*Combining the conventional format with a created format in a memo.*

*Writing the letter of transmittal, abstract, introduction, body, conclusions, recommendations, footnotes, and bibliography of a written report.*

*Using headings and graphic aids in a written report.*

*Using rhetorical patterns to organize the parts of a written report.*

*Reading and analyzing letters and memos written by others.*

So far, we have learned about the basic organizational or rhetorical patterns needed to identify and describe to an audience subjects at rest and in motion. We have also discussed how to convince an audience to accept and act upon our ideas. All these patterns can be combined to create an unlimited number of communications. The audience and purpose of each communication determine which patterns we choose. In this chapter and the next, we study one more factor in planning and creating a communication—format.

The format of a communication is the shape or frame as well as the general organization or arrangement of the communication. Three conventional frames or formats used in business and technical communication are the letter, the memo, and the report. We choose the appropriate conventional format for our communication according to our audience and purpose. The overall organization or arrangement of each conventional format, whether a letter, memo, or report, is based on a combination of rhetorical patterns also appropriate to each specific audience and purpose. In this chapter, we study two aspects of format: the conventional frame or shape of the communication and the overall organization or arrangement of rhetorical patterns.

# Letters

Generally, we write business letters to persuade our audience to take some kind of action: buy our product, pay bills, hire us, and so on. Letters address an audience outside the company, most often the concerned public or our customers. "Customers" may be private individuals or other companies who buy our goods and services. Regardless of the audience, purpose, and subject of the letter, all business letters must accomplish one task: building goodwill. Goodwill is an attitude of support, friendship, and approval—the prestige a company earns that is worth far more than the products it sells. Business letters build goodwill by expressing a genuine "you" attitude.

## The "You" Attitude

The "you" attitude expresses an honest concern for the audience's, rather than the writer's, needs and interests. The "you" then is the audience or recipient of the letter, and business letters express the "you" attitude through style and tone.

As we learned in our discussion of persuasion, we convince audiences to take the action we suggest because that action will benefit them. Of course, we are usually rewarded in some way too, but persuasion as well as goodwill depends on our addressing our audience's needs. For example, when soliciting a customer's order, we would not stress how much we need that order but how much the customer will benefit from the purchase. Of course, we must honestly be interested in how the members of our audience will benefit, or they will distrust our claims. Thus the "you" attitude not only persuades an audience to take the action we suggest, but also, honestly expressed, builds goodwill for our company.

Notice the difference in attitude between the first and second versions of the following collection letters. While the first writer is concerned with his own interests and his attitude may cause ill will, the second writer addresses his audience's interests and builds goodwill:

> We have not received your June payment on your "Preferred Customer" credit account. Late or ignored payments not only hinder our bookkeeping but also cost us time and money for collection services. However, if we do not receive your payment by July 31st, we will be forced to turn your account over to our legal department. To prevent this action, please send your payment promptly.

> We have not received your June payment on your "Preferred Customer" credit account. As a "Preferred Customer," you enjoy the benefits of advanced sales as well as special holiday discounts. You would not want to lose these privileges because of a mislaid or forgotten payment. Please send your payment before July 31st to retain your "Preferred Customer" status.

While the first version attempts to threaten the audience, it is ineffective because it stresses the company's rather than the customer's problems; the second version not only credits the audience with simply forgetting to pay but also provides an incentive to pay promptly. The "you" attitude helps us persuade our audience as well as maintain a friendly atmosphere for the next business transaction.

## Positive and Personal Tone

To support the "you" attitude, most business letters express a positive and personal viewpoint. A positive, confident tone encourages our audience to trust us and to take the action we suggest. For example, instead of saying, "We cannot deliver the shipment until Friday," we can say, "We can deliver the shipment on Friday." Simply stating things positively leaves our audience with the impression that we are in control of the situation. If we do have to report a problem, we can subordinate the negative and stress the positive. For example, instead of saying, "We are behind schedule," we can say, "Although we are behind schedule now, we will meet our final deadline." If we emphasize what we can do rather than what we cannot, the members of our audience will feel that we have their best interests at heart.

Although letters are the most personal form of written communication, unfortunately some writers feel that business letters should be formal and expressed in a business "jargon." This jargon detracts from the "you" attitude by making our company appear large and impersonal. Such expressions as the following detract from a personal tone:

please find enclosed in acknowledgment of
allow me to at the present time
in receipt of as per
in reply to along these lines
for your information in view of the above

A personal conversational tone conveys to the members of the audience that they are dealing with people, not an uncaring organization. The jargon in the following letter hides its "you" attitude as well as makes the letter difficult to understand:

> In acknowledgment of your letter of July 20th in reference to your "Preferred Customer" account, we are hereby crediting your June payment to your account in the amount of $50.75. Please be advised that according to our records your account is paid to date and no interest charge accrued therewith nor customer status changed. We wish to express our appreciation for your prompt action along these lines.

The rewritten version conveys a "you" attitude and a personal tone, and is simple to understand:

> Thank you for your June payment of $50.75. Your account is now up to date. Because you paid so promptly, we have waived all interest charges. Thank you again for being a "Preferred Customer."

A positive and personal tone not only supports the "you" attitude but also helps us write clearly and directly.

## The Conventional Format of a Letter

Most business letters follow a conventional pattern. Because such letters must be clear and direct as well as provoke action, we limit our subjects or topics to one per letter. We also write in short paragraphs, each of which covers one aspect of the subject. Along with limiting the subject and the length of paragraphs, we also divide the letters into three parts: opening, body, and closing.

*The Opening.*   In the first paragraph, we announce our subject as well as attract the audience's attention and convey the "you" attitude. We acknowledge briefly any previous correspondence and state our response to that correspondence. The opening paragraph establishes the tone of the entire letter.

*The Body.*   In the body of the letter, we supply the information the audience requested, solve the problem, show evidence or logical reasoning for our suggestions, explain our decision, and so on. The body of the letter consists of any combination of rhetorical patterns necessary to accomplish our purpose. The content and style of the letter are determined by the specific audience and purpose.

*The Closing.*   The last paragraph in a letter suggests concrete action. In some letters, we also provide an incentive for that action: for example, "If you send in the attached coupon before July 31st, you will receive a free Dentoflex toothbrush." In the last paragraph, we also reaffirm the "you" attitude and goodwill established in the first paragraph. Thus the ending of a letter suggests action, tells the audience how to take this action, if possible provides an incentive for the action, and reestablishes goodwill.

## Combining Conventional and Created Formats in Letters

Although in business letters we follow the conventional format we just discussed, we also "create" a format appropriate to our specific audience and purpose. We create this format by choosing and combining rhetorical or organizational patterns. Let us analyze one type of letter that we all have to write at some time—a letter of application—to see how to combine a conventional format with a created one.

*The Letter of Application.*   The audience of an application letter is a future employer, and the purpose of such a letter is to sell ourselves. The "you" attitude conveys how the employer will benefit from hiring us. In the conventional format, the opening of the letter must attract the employer's attention, the body offer proof that our abilities meet the employer's requirements, and the closing suggest that the employer interview or even hire us. We must be positive about our abilities yet show some of our personality. Since our experience and education determine our abilities, one rhetorical pattern we would use to support the conventional format might be causal analysis. Also we might compare our past experience with the tasks we would have on the job. Certainly the letter should be logical and persuasive. Although we might use other rhetorical patterns to organize the body of such a letter, in the following example the writer depends primarily on causal analysis, comparison, and logical reasoning to sell himself:

75B Main Hall<br>
Potsdam, NY 13676<br>
October 25, 1980

Mr. Richard Brown<br>
Eastman Kodak<br>
1987 Hill Street<br>
Rochester, New York 14626

Dear Mr. Brown:

John G. Phillips has informed me that your firm is looking for a Market Analyst for your West Coast office. My previous work experience and recent college training in the marketing field would enable me to make a substantial contribution to your firm in this position. Please consider me as a candidate for the opening as Market Analyst.

My marketing management program at Clarkson College emphasized the producer-consumer relationship which you state is essential to your marketing program. Also, although my formal marketing training has been good, the key to my education has been diversity. Studying effective business communication skills and the arts has given me a broad outlook toward both the producer and the consumer.

Since your program is designed to provide guidance and assistance to all sales representatives through training programs and customer-information seminars, my previous management positions and my experience working with people give me an edge in this area. These activities have developed my ability to lead and motivate others in a positive direction, a key factor in reaching company goals. For example, as Assistant Building Manager of Center Ice Company, I supervised 15 employees as well as scheduled work activities and collected accounts.

The Marketing Analyst position at Kodak also requires formulation of marketing projects and development of sales-promotion and sales-development plans. Since my work experience at the Center Ice Company required promotion as well as administrative work, the areas you find essential to the job are not new to me; therefore, I think I would be an immediate asset to Kodak.

Along with my work and educational experiences, my extracurricular activities have helped me develop my character. My association with the Big Brother Society has developed my sense of responsibility and given me maturity and independence. These qualities are further exemplified by my being chosen captain of the varsity hockey team in my junior year. The members of the team showed great confidence and trust in my ability to negotiate with management and to deal with people on a personal level.

Since the key factor to a sales and marketing position is the ability to sell yourself, please give me an opportunity to do so in a personal interview at your convenience. I can be reached by phone at (315) 555-8749 or at the address above. I look forward to meeting you.

> Sincerely,
> Bryan Cleaver

This writer uses the conventional format—opening, body, and closing—to sell himself. He uses rhetorical patterns to support his proposition. In his opening, he attracts his audience's attention by mentioning his Kodak contact, he states his purpose, and he emphasizes the "you" attitude. In the body of the letter, he proves by causal analysis and logical reasoning that his abilities match the job requirements. Notice the enthymeme: "Since your program is designed to provide guidance and assistance to all sales representatives through training programs and customer information seminars [premise], my previous management positions and experience working with people give me an edge in this area [conclusion]." In the body, the writer also compares his past experience with the tasks involved at Kodak. Finally, the writer uses induction to prove that his extracurricular activities have developed his character. He closes the letter with a call for action and an incentive to take this action. He ends by reestablishing goodwill.

In any business letter we combine the parts of the conventional format with combinations of rhetorical patterns we choose to meet our specific needs.

# The Resumé

Most letters of application are supported by a resumé or outline of education and work experiences, honors, and personal interests. The letter of application highlights the impressive aspects outlined in the resumé while the facts in the resumé back up the arguments in the letter. The resumé is, of course, much more detailed than the letter and for this reason must be in a very readable or "skimmable" format.

For example, the following resumé accompanied the letter of application we just studied. Notice how the writer selected certain aspects in his resumé, such as his work experience at the Center Ice Company and his work with the Big Brothers and the hockey team, to *prove* he was a good candidate for the job. The rest of the information on the resumé creates the impression that he is a well-rounded, well-educated, experienced individual:

BRYAN L. CLEAVER

| *Permanent Address* | *Current Address* |
|---|---|
| 27 River Lane | 75B Main Hall |
| Toronto, Ontario M3N 1J4 | Potsdam, New York 13676 |
| (416) 555-2793 | (315) 555-8749 |

PROFESSIONAL OBJECTIVE

A position in a firm's marketing department, eventually leading to a marketing research position.

EDUCATION

Bachelor of Science, Management
May 1981
Clarkson College of Technology
Potsdam, New York 13676

Area of Concentration: Marketing Management
Supporting Courses:  Principles of Marketing, Industrial Marketing, Marketing Management, Sales Management, Corporate Finance, Psychology, Principles of Management, Operations Management.
Honors:  Four-year Hockey Scholarship.
Outstanding Athletic and Academic Achievement Award, from the Canadian Government.
Captain of the Varsity Hockey Team.

EXTRACURRICULAR ACTIVITIES

Big Brothers Society of Canada: Helping young boys learn sporting skills and experience life without a father.
Center Ice Sports Corp.: Teaching skating and hockey skills.
Clarkson Varsity Hockey: 4-year letterman.
Intramural Sports: Baseball, Soccer, Golf, Tennis, and Basketball.
New York State Red Cross Blood Drive: Volunteer worker.

WORK EXPERIENCE

| | |
|---|---|
| Assistant Manager | Boiler Room Tavern, Potsdam, New York; Summer 1980. Opened and closed tavern, inventory and cash control. |
| Delivery Salesman | Coca Cola Bottling Company, Toronto, Ontario; Summer 1977–1979. Kept an up-to-date account of deliveries and receipts. |
| Tutor | Clarkson College, Potsdam, New York; Spring and Fall 1980. Instructed Freshman Accounting; Administered and graded tests. |
| Hockey Instructor | Center Ice Company, Toronto, Ontario; Summer and vacations from 1977 through 1980. Instructed skating and hockey; Promotion and administrative work. |
| Assistant Building Manager | Center Ice Company, Toronto, Ontario; Full-time 1975 through 1977. Supervised 15 employees, involving scheduling, public relations, and collection of accounts. |

PERSONAL DATA

Birth Date: April 8, 1956          Marital Status: Single
Height: 5'9"                              Travel: Willing to travel and relocate
Weight: 175 lb                          Date of Availability: June 1981
Health: Excellent

INTERESTS

Camping, reading, music, theater, and most professional sports.

CREDENTIALS

Mr. Jack Newton                           Mr. Len Smith
Varsity Hockey Coach                    General Manager
Potsdam, New York 13676           Center Ice Company
(315) 555-8792                              Toronto 5, Ontario
                                                       (416) 555-7539

Mr. Ralph Nilou
89 Brook Drive
Rochester, New York
(716) 555-6139

The resumé then is a list of accomplishments that are used as supporting evidence for the assertions made in the letter of application. The resumé helps us convince the members of our audience that employing us will be to their benefit.

# Memos

Memorandums or memos, often the main form of written communication *within* an organization, address supervisors, peers, or employees. Three types of memos circulate within an organization: those that announce events, those that function as internal letters, and those that investigate and recommend action. Memos that announce events are brief notices; for example, "All department heads will meet with the Personnel Director to discuss July raises on June 1st, at 2:00 P.M., in room 128." Memos that function as internal letters generally request or pass on information and follow the traditional format of the letter. In this section, we examine memos that are short, informal reports in which we evaluate a situation and recommend action to a supervisor or a peer or in which we direct and explain action to an employee.

## The Conventional Format of a Memo

Since most memos carry a subject line at the top, our audience knows immediately what we are going to discuss. Although less uniform than the conventional format of a letter, memos also have an opening or introduction, a body or discussion, and a closing.

In the opening to a memo, we relate our purpose to the subject announced at the top of the memo. We may also state who authorized us to investigate the problem, the background of the study, our method of investigation, and the order of our discussion within the memo. Since memos often have a wide circulation, sometimes we also state our conclusions and recommendations within the first paragraph for the busy members of the audience. In the body of the memo, we use rhetorical patterns to discuss our findings or reasoning. In the end of the memo, we summarize the discussion or simply restate our conclusions and recommendation.

## Combining Conventional and Created Formats in Memos

While memos must be as positive, concise, and direct as letters, because of the variety of audiences and purposes we address in memos, we depend even more on rhetorical patterns to create our format. The body of a memo that functions as a short report is lengthy and often calls for a greater development or variety of rhetorical patterns. For example, in the following memo, the Director of the Public Works Department writes to the Director of the Highway Department within the same state organization. The writer uses narrative, process description, and deduction to support his recommendation. Because he is writing to a peer, he can use some technical language. He follows the conventional format of a memo as well as creating a format for this particular discussion:

TO: Director, State Highway Department
FROM: Director, Public Works Department
SUBJECT: Drainage System for I-90

In the last few months, we have discussed on many occasions the problem of adequate provision for storm water runoff for the new Interstate Highway I-90 westward to the state line from I-370. As you know, this portion of the interstate system will accumulate a substantial amount of drainage water which will flow into the central industrial system east of the Crow River. As indicated in construction plans, this tributary drainage will be discharged into the present sewer system in the central industrial district. Since this system becomes excessively overloaded with storm water runoff during heavy storms even with the drainage it now accommodates, I propose that we build a new drainage system for I-90 only.

**Background**

Because of my concern about the additional runoff that will be provided by I-90, last May I asked my engineering staff to review alternative methods for disposal of drainage water from I-90. Through May and June, the engineers studied the existing storm sewer that leads to the central industrial district. On July 31, they reported their findings to me: The existing sewer system has the capacity to carry only three to five major storms a year. Since I was sure that this recurrence interval is way below that acceptable for interstate projects, I asked the engineers to investigate the possibility of a new drainage system for I-90 only. Last week they reported their findings.

**Proposed System**

A new system could be placed in the I-90 right-of-way and extend to the Crow River. A 5-by-3-foot concrete pumping station could be placed 200 feet beyond the intersection of Long Avenue and First Street. The pumping station would force the drainage water into the river during high river stages. When the river is low, the drainage water would flow naturally into the river, and the pumping station would not be used. Such a pumping station and drainage system would cost approximately $750,000 but would accommodate 10 major storms a year.

Since the existing central industrial district sewer system is so deficient in carrying present storm water runoffs, I propose that we consider a new system for I-90. If you feel that the system described above is feasible, I propose that our offices request size and cost estimates from local construction firms. Please let me know your reactions to this proposal at next week's meeting.

In the opening of the memo, the writer states the background of the problem, the purpose of the memo, and his recommendation or solution to the problem. He also describes briefly the physical layout. In the body of the memo, he supports his recommendation by narrating his investigation and by describing the "device" and the process involved in draining the runoff. He closes with an enthymeme and a recommendation for definite action.

Memos that function as short, informal reports follow a conventional format of opening, body, and close. The rhetorical patterns we choose to analyze the problem and support the solution depend on the specific audience and purpose of the memo. Had the Director of the Public Works Department wanted to consider equally both the existing drainage system and a new one, he would have compared and contrasted them. However, the Director of the State Highway Department already knew that the old system was inadequate. If the Director of the Public Works Department had the authority to ask for bids on the new system, he would have included specifications in his memo. If his audience had been the mayor rather than another engineer, he would have defined major terms such as tributary drainage in the introduction. However, since his audience was a peer in the same state organization and he wanted to propose a definite system, he narrated what his engineers had found since he last talked with his audience, and then he described the proposed system.

The extent to which we develop the rhetorical patterns we do choose also depends on our specific audience and purpose. Had the Director of the State Highway Department not already known that a drainage problem existed in the central industrial district, the writer would have included more detail in his narrative. Had the writer not already decided on the type of new system, he would have included less detail in his description.

Thus we have as unlimited a variety of combinations and development of rhetorical patterns in memos as we did in letters. We use the conventional format of a memo because it helps us get our message across. The format we create to develop our memo depends on our specific audience and purpose.

# The Conventional Format of a Written Report

Although some reports, such as annual reports to stockholders and consultants' reports to other businesses, circulate outside the company, a written report is usually the most formal communication within an organization and is directed toward a supervisor or boss. We write a formal report for any number of reasons: to propose a project, to report progress, to examine the feasibility of an idea, to report findings from a field trip, to describe periodic profits, and so on. We look at some of these types of reports in the next chapter. The conventional format of any report includes a letter of transmittal, an abstract, an introduction, a discussion, and conclusions and recommendations. Each of the parts meets the needs of the audience and is developed by rhetorical patterns chosen according to the specific purpose of the report. In this chapter, we use a simple business report to illustrate the parts of the report.

## Letter of Transmittal

The letter of transmittal addresses the audience that requested the report. It reminds that audience of the purpose of the project and of the authorization. It may refer to any limitation on the project and describe the methods of investigation. It states the main conclusion or recommendation in the report. Usually clipped to the title page, the letter of transmittal "announces" the report and prepares the audience for what is to come. The letter is more personal and much less technical or complex than any other part of the report.

For example, the following letter of transmittal accompanied a report on a dormitory damage charge system to the newly appointed Dean of Residence Life at a college:

May 12, 1978

Mr. Joseph Dodds
Dean of Residence Life
Main College
Mainline, New York 13700

Here is the report you authorized on April 5th for an analysis of the current dormitory damage charge system used at this college.

The purpose of this report was to determine the effectiveness of the system and to offer any concrete recommendations for improvement. To do this, I analyzed in detail the damage costs figures for the past 3 years and did an extensive study of dormitory conditions. Although I had limited manpower, I gathered information on all seven dormitories.

My immediate findings indicate that the system is functioning well. However, improvements on the physical characteristics of the dormitories, such as new carpeting and paint, would make the system even more efficient.

I have enjoyed conducting this study and hope my findings help you make your final decision. If I can be of any further assistance, please contact me.

Sincerely,<br>Barbara Strollo<br>Research Analyst

Because the writer states her conclusions, the audience can anticipate the evidence that must follow.

## The Abstract

Abstracts are brief summaries of a formal report. Often we receive a great many reports within an organization. We read in detail those we have authorized or those that affect our jobs; for those that we are too busy to read or that do not concern us directly, we depend on the abstract to give us an idea of the whole report. Since everyone who receives a copy of the report reads at least the abstract, it must be brief, concise, and usually less technical than the report itself. In an abstract, we state the purpose of our report, the problem we investigated, the methods we used, the key ideas within the report, the conclusions we reached, and the action we recommend. The abstract is more detailed and complex than the letter of transmittal but less so than the introduction to the report; for example, the abstract of the report on the dormitory damage charge system follows:

The dormitory damage charge system has existed since 1975. The purpose of this report was to determine the effectiveness of the system and to offer any suggestions for improvement. Research methods for the study included a detailed analysis of the damage costs incurred since the system's initiation. After careful analysis of these costs, I conclude that the system should be kept in operation.

Study of individual dormitory costs revealed that certain factors contribute to the high incidence of damage in H. House, C. House, and M. House. Damage costs for these three complexes accounted for 80 percent of total costs charged to all dormitories last semester. These contributing factors include: the shabbiness of the interior decor, the segregation of upper- and underclassmen, and the absence of women residents.

To further decrease annual damage costs, I have several suggestions for improvement of conditions in these three dorms; these suggestions include laying down new carpets, installing more lights, resurfacing walls, integrating freshmen, sophomores, juniors, and seniors, and placing women residents in what are now all-male dormitories.

Anyone reading this abstract would know the problem the writer investigated, her methods, her key ideas, and her conclusions and recommendations.

## The Introduction

The introduction to a report gives all the background information that the audience needs to understand the report. While some members of the audience may stop with the abstract, anyone who reads the introduction will probably read the whole report. The introduction then serves as a bridge into the report. Introductions include the history of the situation or problem, the origin or authorization of the report, the purpose of the report, the scope and exact coverage, any limitations or impediments to the research, definitions, sources and methods of collecting data, and a preview of the report or order of presentation. The introduction is more detailed than the abstract and contains anything that will help the audience understand what is to come in the report.

The following introduction contains a brief narrative of the history of the subject, an explanation of the procedure now used, a contrast of the present and past procedures, an explanation of methods, and a brief preview of what will follow. The writer uses rhetorical patterns here that reappear in the discussion section or the body of the report:

### Introduction

Dormitory damage fees are considered by most administrators to be a necessary cost in operating a college today. In an attempt to keep these costs to a minimum, in the past few years resident life directors have adopted several systems to deal with the problem of dormitory damage and breakage.

This report is an analytical study of the dormitory damage charge system at this college. The system is designed to place the responsibility of the damage problem directly on the students. Individual damage inquiry forms are posted to inform the residents of the specific damage committed and the cost it will bring to the college. If the individual who committed the damage does not come forward and admit it, the cost is then charged to the residents of the floor or dormitory in which it occurred. The cost is divided equally among the residents of the floor or dormitory and is subtracted from a mandatory $50 damage fee submitted by all students in their freshman year.

This system has been in effect for 3 years. Before that time, no records were kept of dormitory damages. The purpose of this report is to determine whether the system is working well enough to be kept in operation and to offer sound recommendations for any improvement.

Research for the report consisted of detailed analysis and comparison of damage costs for each floor and dormitory over a period of the past 3 years. Figures for the spring 1978 semester were not available at the time this report was written. All tables from which the information was extracted appear in the appendixes, along with a copy of the damage inquiry form.

The results of this comparative analysis are detailed in the report.

Because the success of the damage charge system depends on the methods that are used, we can anticipate that the writer will use the rhetorical pattern of procedure along with her comparison-contrast. The introduction not only introduces the audience to the content of the report but also often indicates what rhetorical patterns are used to organize the content.

## The Discussion: Mechanical Elements

We will now study how to combine conventional and created formats because the discussion section, or body of a report, consists more of combinations of rhetorical patterns than of conventional parts. While some reports are organized chronologically or developmentally, inductively or deductively, there is no set format for a discussion section. Most writers examine the problem and then analyze possible solutions, but again we can use an unlimited variety of rhetorical patterns in these discussions. Before examining the discussion of our report on dormitory damage, we need to study two mechanical aspects of the discussion: headings and graphic aids.

**Headings.**   To break up long sections of writing and to help our audience keep track of the arrangement and direction of our report, we label the various sections of the discussion. Headings can come from our outline and may be listed in a Table of Contents preceding the abstract. Our audience will understand the discussion more easily if we indicate in headings the subject of each section; an audience will read our discussion more carefully if we use headings to divide the prose into manageable chunks. Numbering headings also lets us refer back to sections.

We can "design" headings on the typewriter simply by capitalizing and underlining. The following list shows one system of headings:

# I. FIRST-LEVEL HEADING

## A. Second-Level Heading

The prose would appear below the centered first-level heading and below the second-level heading, which runs to the edge of the left-hand margin.

### 1. Third-Level Heading.

The third-level heading begins a paragraph, and each word of the heading begins with a capital.

a. *Fourth-level heading.* The fourth-level heading begins a paragraph, but only the first letter in the heading is capitalized, and the whole heading is underlined or italicized.

No matter what system or design we use, we should divide our discussion by headings.

**Graphic Aids.** In earlier chapters, we discussed graphic aids as appropriate to particular rhetorical patterns. Since we use combinations of rhetorical patterns in reports, we choose various graphic aids. Our choice depends on our audience, purpose, and rhetorical patterns. If the audience is a nontechnical one, we can use simple graphic aids, but we must explain and evaluate the data thoroughly in our discussion. If the audience is a technical one, experienced in reading graphic aids, we can depend on more complex aids and less written explanation. In either case, graphic aids supplement our writing rather than take its place.

Graphic aids appear either immediately after or near our explanation if they are essential to the discussion. If they provide additional data that only some of our audience will be interested in, we can place them in an appendix to our report. However, all graphic aids should be labeled according to type, number, and caption (e.g., Table 1 Dormitory Damage 1975 or Figure 7.6 Automobile Sales 1979–1980). Anything other than a table, such as a diagram, line graph, or photograph, is labeled a figure. We should refer to graphic aids in our discussion to alert the audience to them (e.g., see Table 1 or see Figure 7.6).

**Tables.** Tables are most useful in comparison-contrast, causal analysis, and classification when we present many specific facts and variables for rapid assimilation. The major headings of a table are represented in Table 13-1.

**TABLE 13-1.** Title

| Stub Head | Column Head | Column Head (unit of measurement) | |
| --- | --- | --- | --- |
| | | Subhead | Subhead |
| Line head | Data | Data | Data |
| Subhead[a] | Data | Data | Data |
| Subhead | Data | Data | . . . or NA[b] |

[a]Footnote.
[b]Not available.

*Source:* Author, title, volume and date, page number.

Stub (the main subject of the table), column, and line heads identify the name of the items compared, and the unit of measurement (such as inches, years, or dollars) appears under the column head when possible. Column and line heads should be ordered by time, quantity, or alphabetical order. A simple table can appear in the text of a report while a complicated or supplementary table would appear alone on a page or in an appendix. The more complicated the table, the more the audience needs footnotes and lines to identify and separate the material (see Table 13-2).

**TABLE 13-2. Maximum Process Fluid Velocities\* for Thermowells**

| Thermo-well Size | Tempera-ture Bulb Size In. | mm | Material | Insertion Length† In. (mm) | | | | | | | |
|---|---|---|---|---|---|---|---|---|---|---|---|
| | | | | 7.5 (191) | | 10.5 (267) | | 16 (406) | | 24 (610) | |
| | | | | Ft/s | m/s | Ft/s | m/s | Ft/s | m/s | Ft/s | m/s |
| ½ NPT | 0.38 | 9.2 | Brass | 38 | 11.6 | 19 | 5.8 | 8 | 2.4 | ... | ... |
| | | | Carbon Steel | 48 | 14.6 | 25 | 7.6 | 11 | 3.4 | ... | ... |
| | | | 304 SST/316 SST | 50 | 15.2 | 26 | 7.9 | 11 | 3.4 | ... | ... |
| | | | Monel | 48 | 14.6 | 24 | 7.3 | 11 | 3.4 | ... | ... |
| ¾ NPT | 0.38 | 9.2 | Brass | 54 | 16.5 | 27 | 8.2 | 12 | 3.7 | ... | ... |
| | | | Carbon Steel | 69 | 21.0 | 35 | 10.7 | 15 | 4.6 | ... | ... |
| | | | 304 SST/316 SST | 72 | 21.9 | 37 | 11.3 | 16 | 4.9 | ... | ... |
| | | | Monel | 68 | 20.7 | 35 | 10.7 | 15 | 4.6 | ... | ... |
| ¾ NPT | 0.56 | 14.3 | Carbon Steel | 97 | 29.6 | 49 | 14.9 | 21 | 6.4 | 10 | 3.0 |
| | | | 304 SST/316 SST | 100 | 30.5 | 51 | 15.5 | 22 | 6.7 | 10 | 3.0 |
| | | | Monel | 95 | 28.9 | 49 | 14.9 | 21 | 6.4 | 9 | 2.7 |

\*For gas, air, or steam. Values may be lower for liquids.
†This is the "U" dimension in figure 6.

*Line Graphs.* Line graphs are most useful in representing continuous relationships or trends such as in narrative and process description or any pattern concerned with time or distance. The independent variable such as time usually appears on the horizontal axis, and the dependent variable on the vertical axis; a line is drawn between the plotted data items to show trends. We can differentiate between various relationships by using solid or dotted lines on the graph, but more than three lines may make the graph difficult to read. A simple line graph (such as Fig. 13-1) allows the audience to gauge the increase in the differential gap or percentage of range on the temperature controller as the setting is turned up. In the complex line graph (see Fig. 13-2) patterns such as dots and dashes help the audience distinguish between the lines on the graph. A key appears on the graph. The scale on the graph should not exaggerate or minimize the points.

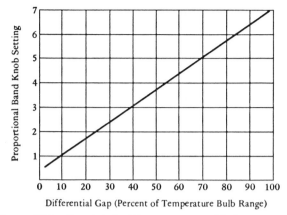

Figure 13-1. Reprinted with permission of Fisher Controls Company, Marshallton, Iowa.

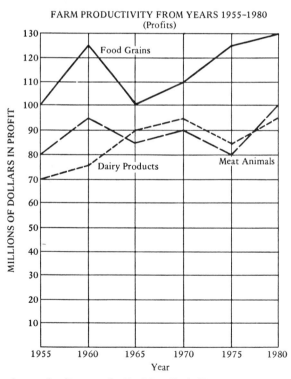

Figure 13-2. A complex line graph. By: Mary Beth Hagen

**Bar Graphs.** Bar graphs show relationships or trends that are not continuous, such as those in comparison-contrast or processes. Usually the variable expressing time or distance appears on the horizontal axis, and the variable expressing quantity on the vertical axis. The bars on bar charts are often divided so that the audience can see not only the relative quantities but also the segments of those relative quantities. To clarify a bar graph we often note the percentage of or quantity over each bar. The bars on the graph must be spaced so that the audience can make an easy comparison, but so that the bars are not crowded. The height of the bars should show—at a glance—the increase or decrease, growth or decline of an item. Usually the bars are vertical except when such things as distance are compared. The bar graph is highly visual, but the audience may not be able to see precise figures as easily as on a line graph or table. A combination bar and line graph (Fig. 13-3) can be used to show both trend and comparison.

**Circle Graphs.** Circle graphs or pie charts represent symbolically the percentages or components of a whole or 100% of a subject at a set time. They aid in any comparison-contrast or partition for a nontechnical audience. Our labeling, rather than the circle chart itself, indicates exact figures. In drawing a circle graph, use a compass to draw the circle and start the largest segment of the divided circle at the 12:00 position and proceed around the clock to the

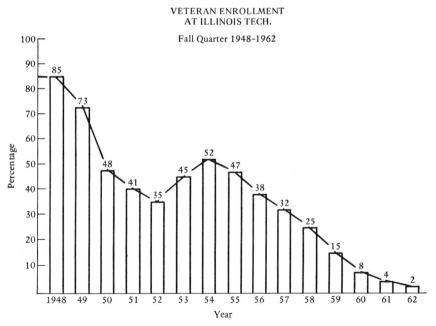

Source: Registrar's Office

*Figure 13-3.* Combination bar-and-line graph by Andrew T. Miller.

smallest segment. Usually a circle graph contains between three and eight segments, the smallest segment perhaps labeled "other." To calculate how much of the circle to leave for each segment, first calculate the percentage of each segment. Remember that a circle contains 360 degrees; if your segment comprises 60% of the whole (100%), then the portion of the circle allowed for that segment can be calculated as follows: 60%/100% = x/360 degrees or x = 216 degrees of the circle. Again segments should be labeled with the name of the item and the percentage so that the audience can both visualize the proportions and read the exact percentages. The portions can also be colored or marked to distinguish them (see Fig. 13-4).

**Pictographs.**  Pictographs are graphs that use an image or symbol to represent the item within the graph. They are most useful in presenting multiple subjects to a very nontechnical audience, an audience not used to reading other kinds of graphic aids, or an audience not willing to give much time to graphs. Pictographs are not only visual but precise if we indicate exact quantities next to the symbols. The symbols are simplified drawings called "glyphs" or "isotypes," have the same widths, and are aligned vertically and horizontally in single rows. The glyphs or isotypes should be self-explanatory drawings representing approximate quantities of one whole or one half, seldom less (see Fig. 13-5).

**Diagrams.**  Diagrams are essential to device and process description, instructions, and specifications. Each part of the device must be labeled by name, and motion indicated by arrows. Technical audiences are usually well experienced in reading diagrams. A photograph of the device is often used in place of a diagram to lend realism to the illustration and add shades of gray for depth.

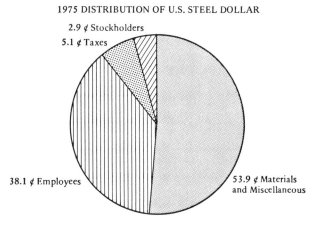

1975 DISTRIBUTION OF U.S. STEEL DOLLAR

2.9 ¢ Stockholders
5.1 ¢ Taxes
38.1 ¢ Employees
53.9 ¢ Materials and Miscellaneous

Source: 1975 Stockholder Bulletin

*Figure 13-4.*  Pie chart by Mary Beth Hagen.

FARM PRODUCTIVITY (PROFITS)

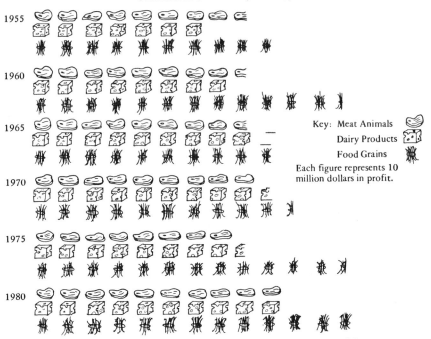

*Figure 13-5.* Pictograph by Mary J. Schaad. *Source:* U.S. Bureau of Census.

**Cutaway Diagrams.** Cutaway diagrams "remove" the device's casing to show the inside of the device, the relationships between the inner parts to each other and to the whole. Cutaway diagrams show what the naked eye cannot see or what the repairman *will* see upon opening the device; however, since most devices are taken apart piece by piece, a cutaway diagram shows the interior of the device as it would look still together. Figure 13-6 illustrates the inside of an instrument case terminal; the jagged edge indicates to the audience that the exterior is "cutaway." Figure 13-7 is a more complicated cutaway diagram of a valve body. Notice that all parts are clearly labeled and the exterior remains a shaded outline.

**Exploded Diagrams.** In an exploded diagram the parts seem to be "blown" apart, yet the parts appear in their normal arrangement. Such diagrams are useful in supporting device description or instructions. The audience can see each part as a whole and as related to other parts. Figure 13-8 is an exploded diagram of a valve and its parts.

**Process Diagrams.** Process diagrams can be cutaway diagrams with arrows to show the direction and movement of each part and support process descriptions. These diagrams can also represent the flow or movement between whole devices or parts of devices

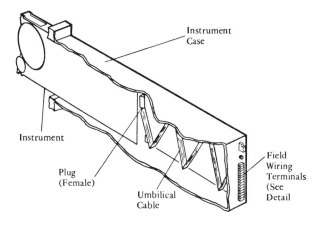

FIELD WIRING TERMINALS
ON TYPICAL ac² INSTRUMENT CASES

*Figure 13-6.* Instrument case terminals. Reprinted with the permission of Fisher Controls Company, Marshalltown, Iowa.

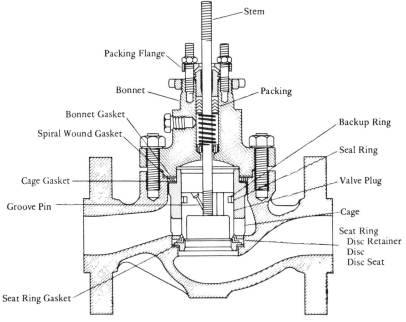

*Figure 13-7.* Sectional view of design ET valve body with full-size trim. Reprinted with the permission of Fisher Controls Company, Marshalltown, Iowa.

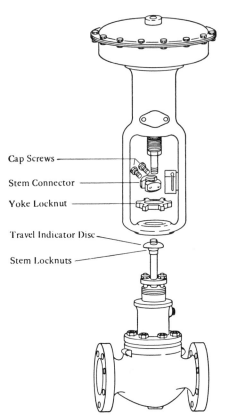

Cap Screws

Stem Connector

Yoke Locknut

Travel Indicator Disc

Stem Locknuts

*Figure 13-8.* Actuator mounting (type 657 actuator shown). Reprinted with the permission of Fisher Controls Company, Marshalltown, Iowa.

represented only by boxes or circles. Process diagrams help an audience visualize what a device looks like in action or the chronology between devices in a process. We saw process diagrams in Chapter 9. Figure 13-9 illustrates a process description for an audience not interested as much in what the device looks like but in the chronology of the process; thus the parts of the device as represented by boxes. For the technician who must repair and troubleshoot a complex device, more complicated diagrams are essential. A schematic diagram (see Fig. 13-10) is really a complicated process diagram or a flowchart; the parts of the device are represented rather than drawn, but a great deal of information is given in each box or circle.

Again we choose the type of graphic aid according to our audience, purpose, and rhetorical pattern. In the discussion section of a report, we can use any variety of graphic aids.

## The Discussion: An Example of a Created Format

The writer of the report on the dormitory damage system organizes her discussion section by combining four rhetorical patterns: procedure, comparison-contrast, causal analysis, and description. Her

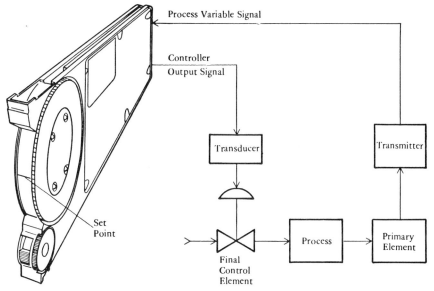

*Figure 13-9.* Simplified process control loop. Reprinted with the permission of Fisher Controls Company, Marshalltown, Iowa.

development of these patterns depends on her audience and purpose. Since her audience must understand thoroughly how the system works, she elaborates on the procedure noted in her introduction. She lists all possible alternatives or options in that procedure. Then, in a line graph she shows the trend in dormitory damage costs for 3 years and in a table gives the exact figures for each dormitory. These graphic aids allow her audience to compare trends as well as statistics. Finally, she describes the major characteristics. These rhetorical patterns allow the audience to understand the procedure, compare costs, visualize the dormitories, and anticipate the changes the writer recommends:

### Explanation of the System

The dormitory damage charge system at Clarkson involves the use of individual damage inquiry forms that are filled out by the residence life director and posted in the respective dormitories whenever an act of damage is reported.

The purpose of these forms is to communicate to students the act of damage that has been committed and the cost that it will bring to the college. Ultimately, the goal of the system is to charge only those individuals who are responsible for the damage. To accomplish this, the system does not in any way threaten those who admit to committing the damage. There is no punishment, no penalty, no fine. The cost is simply subtracted from the student's $50 damage deposit which he or she submitted during the freshman year.

If a student is turned in by another student, the student is entitled to an appeal. He or she must come before the residence life staff and cite reasons for innocence. No student is charged without an appeal process.

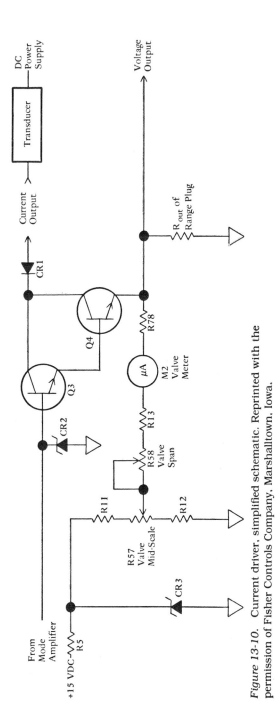

*Figure 13-10.* Current driver, simplified schematic. Reprinted with the permission of Fisher Controls Company, Marshalltown, Iowa.

If no student admits to the damage and no student is turned in by another student, the cost of the damage is charged to the residents of the floor or dormitory where it occurred. Damage committed on a specific floor is charged to the residents of that floor. Damage committed in a common area of the dormitory, such as main lounge or community bathroom, is charged to all the residents of the dormitory. Costs are divided equally and subtracted from each student's $50 deposit.

By making students aware of the damage problem, the floor-and-dormitory charge system strives to encourage students to be responsible for their own actions. In this way, it is hoped that students will take preventative measures and that the overall yearly damage costs will go down.

### Comparison of Total Yearly Costs

Although the yearly damage costs do not include the figures for the spring semester 1978, they indicate that damage costs for the academic year 1977–1978 will decrease from costs of the previous year (see Fig. 13-11). These figures reflect damage in 16 campus dormitories.

Costs for the fall semester 1977 equalled only one-third of the total costs for the previous year. It is unlikely that costs incurred during the spring semester 1978 could cause the total costs for the academic year 1977–1978 to surpass those of the previous year.

The high costs of the academic year 1976–1977 were due in part to one particular incident in which residents from C. House bombarded the windows of the H. House with snowballs. This incident resulted in $4,000 worth of damage in 20 minutes.

Costs charged to individuals were nearly equal to those charged to floors and dormitories. This fact indicates that approximately half of the damage charges could not be attributed to the individuals who committed the acts.

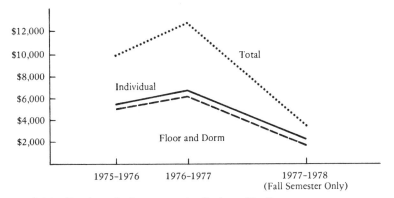

*Figure 13-11.* Total yearly damage costs. Barbara Strollo.

### Comparison of Individual Dormitory Costs

The figures represented in Table 13-3 show several inconsistencies in the damage costs for individual dormitories. Of the 16 dormitories on campus, I concentrated on three "problem" dormitories, H. House, C. House, and M. House, and three contrasting dormitories, G. Hall, P. Hall, and H. Hall.

Costs for G. Hall, H. Hall, and H. House show a steady decrease in costs, while costs for M. House and C. House show a steady increase. Costs for P. Hall appear to be remaining relatively stable.

**TABLE 13-3.   Total Individual Floor-and-Dormitory Charges**

|  | 1975–1976 | 1976–1977 | 1977–1978 (Fall Semester Only) |
|---|---|---|---|
| H. House | $800 | $  500 | $240 |
| C. House | 700 | 1,500 | 760 |
| G. Hall | 650 | 220 | 100 |
| P. Hall | 300 | 200 | 180 |
| M. House | 200 | 500 | 540 |
| H. Hall | 20 | 20 | 0 |

### Characteristics of Individual Dormitories

The figures represented in Table 13-3 do not take on any significance until the characteristics of each dormitory represented are taken into consideration.

M. House, H. House, and C. House are all used to house freshmen and sophomores. Although costs for H. House appear to be decreasing, these three dormitories still have the highest damage costs of all the dormitories on campus, accounting for nearly 80 percent of the total floor-and-dormitory costs charged last semester.

These three dormitory complexes are the oldest on campus. Walls are built of cinderblock, halls are very dimly lit, and carpets are old and torn. Residents live two to a room and share a large community bathroom located on each floor.

These three complexes also house nearly all males. However, two floors in C. House house only female residents. It should be noted that these floors had nearly 50 percent less damage charges than the floors which housed all males.

P. and G. Halls are used to house juniors and seniors. These two complexes are the newest on campus. Walls are made of plasterboard, carpets are brightly colored, and each floor has its own central lounge where residents may gather to relax and converse. The dorms are coed, with women residents being randomly dispersed throughout. Residents live in suites—four students in two rooms, with an adjoining bathroom in between. These two complexes accounted for 20 percent of the total floor-and-dormitory costs charged last semester.

H. Hall is a very richly decorated converted mansion and has 12 large rooms which house two residents each. All residents in this House are female. H. Hall had no damage charges last semester.

The amount of detail and the development of each rhetorical pattern depend on the audience and purpose of the report. While the writer included detail about the procedure, she did not have to include all the dimensions and specifications of the dormitories.

## The Conclusions and Recommendations

In this final section of a formal report, we must logically induce or deduce an inference or assertion about the situation we discussed. Then we must propose specific actions. Since we have already stated our conclusions and recommendations in our letter of transmittal and our abstract, our findings come as no surprise to our audience; however, by the end of our report, our audience should agree with our reasoning. Thus, our conclusions and recommendations are persuasive. Note that the writer of our sample report ends with logical conclusions and concrete recommendations:

> The damage charge system is accomplishing its purpose of keeping damage costs down. Although it has been in effect for only 3 years, the system has resulted in some decreases in damage costs or has at least kept them stable. Therefore, the system should be continued.
>
> Several factors contribute to the high costs of damage in certain dormitories. These factors include: the shabby decor, the segregation of upper- and underclassmen, and the absence of women residents. The dormitories where I found these characteristics were M. House, H. House, and C. House. These dormitories accounted for nearly 80 percent of the damage costs charged last semester.
>
> When contrasted with these high costs, the low damage costs in the more modern dormitories seem to indicate that where students appreciate and respect the area in which they live, they do less damage. Taking this into account, I suggest that the conditions in M. House, H. House, and C. House be improved. An improvement in the conditions in these dormitories should result in a higher degree of efficiency of the current damage charge system.
>
> Specific recommendations follow:
>
> 1. Lay new carpets in H. House and C. House.
> 2. Install more lights in M. House, H. House, and C. House.
> 3. Resurface the cinderblock walls in M. House, H. House, and C. House with paneling or plasterboard.
> 4. Assign freshmen, sophomores, juniors, and seniors to all dormitories.
> 5. Place women residents on one floor in H. House, or assign women residents to one-half of each of the floors in both H. House and C. House.
>
> *Barbara Strollo*

We combine the conventional format of a report—the letter of transmittal, the abstract, the introduction, the discussion, and the conclusions and recommendations—with the format we create to suit our specific audience and purpose. While the conventional parts of a formal report seldom vary, our choices and combinations of rhetorical patterns are unlimited. By knowing thoroughly the uses and techniques of each rhetorical pattern as presented in this book, we can choose the detail and development we need to use rhetorical patterns to arrange and organize our discussion.

## Footnotes

Since for many reports we do research in secondary sources, we need to credit other authors with their ideas and statements, usually in the form of a footnote. Often footnotes appear at the bottom of the page on which the idea or statement to which they refer appears, although sometimes they can be gathered at the end of a report in a "Notes" section. In either case, whenever we use the specific words of another author or adopt an idea original to another author, we need to cite our source in a footnote.

Although footnote form may vary slightly in specific fields, the list below represents standard footnote form:

1. Isaac Asimov, *Adding a Dimension: Seventeen Essays on the History of Science* (New York: Doubleday, 1964), p. 4.

2. Geoffrey Norman, "The Satellite That Invaded a Campsite," *Esquire*, Vol. 89, No. 4 (March 14, 1978), pp. 85–86.

3. Norman, p. 87. [This footnote would refer to the one above it; only the page referred to in the source has changed. At one time a footnote referring to the one immediately above would read ibid., p. 87, and a footnote referring to one elsewhere in the footnote list would read Norman, op. cit., p. 93. However, now most people accept the author's surname, the page number, and, if we cite other works by the same author, the short title of the work (e.g., Norman, "Satellite," p. 93, or Asimov, *Dimension*, p. 5) as acceptable form wherever the second reference appears in our list.]

4. John Read, "Chemistry," *What Is Science? Twelve Eminent Scientists and Philosophers Explain Their Various Fields to the Layman*, James R. Newman, Ed. (New York: Simon & Schuster, 1955), pp. 145–155.

5. Fisher Controls Company, *General Catalog 501*, 1975, p. 17.

6. U.S. Department of Health, Education, and Welfare, *Guide for the Care and Use of Laboratory Animals* (Washington, D.C.: U.S. Government Printing Office, 1978), p. 3.

7. Forest Ray Moulton and Justus J. Schifferes, Eds., *The Autobiography of Science*, 2d ed. (New York: Doubleday, 1960), p. 678.

8. "Crime in the Streets," *The New York Times*, September 7, 1980, p. 23.

Footnotes can contain material other than the sources we used in our research. We can also use footnotes for the following purposes:

1. Additional information that is relevant to our report but not essential to every reader.
2. Statistics that support our comments but that every reader need not interpret.
3. Sources that a reader may want to investigate if he or she is interested in doing more research on the subject.
4. Different points of view that we may disagree with but should acknowledge.

In footnotes below our text or gathered at the end of our report, we cite sources or offer additional information not needed by every reader of the report.

## Bibliography

At the end of the report, we often include a bibliography that contains all the sources used in preparing the report, whether we cited these sources in footnotes or not. The bibliography is in alphabetical order as in the following:

Asimov, Isaac. *Adding a Dimension: Seventeen Essays on the History of Science.* New York: Doubleday, 1964.

"Crime in the Streets." *The New York Times,* September 7, 1980.

Fisher Controls Company. *General Catalog 501,* 1975.

Moulton, Forest Ray, and Justus J. Schifferes, Eds. *The Autobiography of Science.* 2d ed. New York: Doubleday, 1960.

Norman, Geoffrey. "The Satellite That Invaded a Campsite." *Esquire,* Vol. 89, No. 4 (March 14, 1978).

Read, John. "Chemistry." In *What Is Science? Twelve Eminent Scientists and Philosophers Explain Their Various Fields to the Layman.* Edited by James R. Newman. New York: Simon & Schuster, 1955.

U.S. Department of Health, Education, and Welfare. *Guide for the Care and Use of Laboratory Animals.* Washington, D.C.: U.S. Government Printing Office, 1978.

The bibliography and, if used, the "note" section appear after our Conclusions and Recommendations.

# Summary

Letters, memos, and reports all employ conventional formats. However, we develop and support the parts of these conventional formats by created combinations and arrangements of organizational patterns.

Letters should express a "you" attitude to gain goodwill and to persuade the audience to take recommended action. In letters we are personal and positive and we try to avoid business jargon. In the opening of a letter, we attract the audience's attention and establish our "you" attitude; in the body we use rhetorical patterns appropriate to our audience and purpose; and in the closing we suggest definite action and reestablish goodwill. We use the conventional format and combinations of rhetorical patterns to organize memos. Most letters are persuasive in some way, such as the letter of application supported by the resumé.

The formal report contains a letter of transmittal, an abstract, an introduction, a discussion, and recommendations and conclusions. As we move through the report, each of these parts is more complex and technical, the letter being most personal and nontechnical, the discussion being most complex. More of the members of our audience read the abstract of a report than any other part. The introduction acts as a bridge into the discussion of the report, which is in turn divided into headings and contains graphic aids. We can organize our discussion according to an unlimited number of combinations of rhetorical patterns. Our choice of patterns and the extent to which we develop them depend on our audience and purpose.

While letters, memos, and reports follow conventional formats in business and technical communication, we support and develop these formats by organizational or rhetorical patterns appropriate to our audience and purpose. We look at reports more closely in the next chapter.

# Reading
## to Analyze and Discuss

### I

April 27, 1980

Mr. Peter Brooke
Brooke and Sons, Inc.
Main Street
Kansas City, Missouri 64130

Dear Mr. Brooke:

As soon as I reviewed your letter of March 19, 1979, about the condition of Main Street from First to Fifth Avenue, I had both our Street Maintenance and Engineering personnel review the area.

The problem stems from the fact that Main is an unimproved street (not surfaced) and as such is not designed to accommodate the loads placed on it by the many trucks traveling on it. This condition combined with the lack of drainage create the situation you currently have.

As a partial remedy, in April the Street Maintenance Section will construct roadside ditches to accommodate the storm water. Adjacent property owners will be required to furnish city specified driveway pipes for their entrances. These pipes will be installed at no additional cost. This process will eliminate roadside parking on the side of the street where the ditches are located.

A more permanent solution would be to construct an improved street capable of carrying the weight of the trucks. This project would include storm drainage improvements and would be at the expense of the adjoining property owners. Should you wish to pursue this alternative, please contact immediately Mr. John Snow, Associate City Engineer, City Hall, Kansas City, Missouri. Thank you for calling this problem to our attention.

Sincerely,
Raymond Caldwell
Director of Public Works

## Questions for Discussion

1. Does the letter use the conventional format for business letters?

2. What rhetorical patterns does the writer use to supplement the conventional format?

3. Does the letter convey a "you" attitude?

4. Is the letter personal, positive, and concise?

# Exercises and Assignments

**1. Problem:** Letters of application are either solicited (responding to a specific opening) or unsolicited (querying a company about any possible openings). Letters of application are developed by both conventional and created formats as we saw in this chapter.

Assignment:

1. Find an advertisement in your local newspaper or at your college placement office for a job that you are qualified to fill. Write a letter of application using a conventional and created format appropriate to your audience and purpose. Enclose a resumé.
2. Write a letter of application to a company you would be interested in working for. Assume that one of many jobs may open up in the future. Use a combination of conventional and created formats appropriate to your audience and purpose. Enclose a resumé.

**2. Problem:** You manage an Italian restaurant in a small town. Recently you have had to raise your prices on all poultry entrees. The local Chamber of Commerce has a banquet at your restaurant the first Monday night of every month. For years, one of the choices of entrees for their banquet has been Chicken Kiev. Recently the chairman of the group has written a letter of complaint about the price increase.

Assignment:

**Write a letter to the chairman explaining or justifying the price increase.**

**3. Problem:** Consider a problem that you observe on your campus, such as long registration lines, poor food service, or book costs. Choose a narrow topic that you could investigate and report on in a week's time.

Assignment:

**Write a memo report to an appropriate audience in which you assess the problem, analyze the causes, and suggest a solution.**

**4. Problem:** Choosing the best graphic aid for a communication is a matter of audience and purpose analysis. Display each set of statistics given below in the best graphic aid for the given audience and purpose.

Assignment:

1. For the annual report (yearly financial communication to stockholders) of a department store: For the Men's Department, Location A, the total sales profit was $10,546 and for the House-

ware's Department, Location A, the total sales profit was $15,687. But in the Men's Department, Location B, the total sales profit was only $5,669. In the Women's Department, Location B, the total sales profit was $4,587, and Location A, $6,872. In the Houseware's Department, Location B, the total sales profit was $9,078.

2. Patient Jones' blood pressure on one day read 135/80 in the morning and 140/85 in the afternoon. Before lunch it read 130/70 and after lunch 145/82. Just before dinner his blood pressure dropped to 125/68 but before bed again read 135/80. In a report to the consulting physician, an intern needs to display and analyze any patterns in Jones' blood pressure readings. Choose and draw the graphic aid.

3. In a presentation to school children the uses of coal in the 1800s are explained. During the 1800s out of every 100 tons of coal consumed, 30 tons were used by industry while 55 tons were consumed in private homes. Out of every 100 tons, 12 tons were sold outside the country and 3 tons were used for other purposes such as in manufacturing other sources of fuel. In the 20th century, out of every 100 tons consumed, 6 tons are used for industrial purposes, 30 tons for heating private homes, and 5 tons for shipment elsewhere or in fuel manufacturing. Choose and draw the graphic aid.

4. For an incoming freshman, graphically depict the way to register for courses at your school.

5. Graphically depict for the Board of Trustees of your school how many students major in the various courses of study offered. (Your college catalog is a good source of statistics.) Divide your findings into the various levels of students: freshman, sophomore, junior, senior, graduate student.

6. Reread the report on dormitory damage included in this chapter and design additional graphic aids to include in the report. You may add or extrapolate any statistics you wish.

# 14

# Common Types of Reports:

## The Analytical Report, the Progress Report, and the Oral Report

*Presenting the advantages and disadvantages of a project or solution.*
*Analyzing and interpreting a situation.*
*Assessing progress up to now and predicting progress in the future.*
*Offering tentative solutions.*
*Delivering a report orally.*
*Helping our listening audience remember, understand, and appreciate our ideas.*
*Reading and analyzing reports written by others.*

As we saw in the last chapter, the conventional format of a report includes a letter of transmittal, an abstract, an introduction, the body, and conclusions and recommendations. In addition, we combine the organizational or rhetorical patterns we have studied throughout this book to create a communication appropriate for our audience and purpose. Now we continue our discussion with a study of three common types of reports: the analytical report, the progress report, and the oral report.

The analytical report investigates a problem or a subject and reveals the essential characteristics of the subject before making a recommendation. Usually in an analytical report we weigh the advantages and disadvantages of our solution or recommendations. We take the audience through our assessment of the problem, explain our methods of analysis, and lead the audience to our conclusions.

A progress report relates what steps have been taken in an ongoing project over a certain period of time. In a progress report, we sometimes redefine our project or problem, give what evidence or information we have gathered so far, and recommend how we should complete the project.

Finally, an oral report can be an oral presentation of a written report or a presentation of information gathered only for the purposes of our speech. Because the members of our audience must listen to rather than read our communication, we must give them special help in remembering and understanding our ideas.

# The Analytical Report

One of the most common types of written reports is the analytical report. We have seen various kinds of analysis already in this book, in particular, process description or analysis and causal analysis. Again analysis is the technique of breaking a subject into its parts to see how these parts contribute to the whole process or effect. In an analytical report, we analyze the subject, often a problem or a new procedure, and its significant parts to see and weigh characteristics and to recommend a solution or an approach.

For example, in the following report, the writer analyzes plasma arc cutting. Although he uses definition, description, and comparison-contrast, he primarily analyzes the advantages and disadvantages of this new process. After his analysis, he concludes that despite certain disadvantages, the plasma method saves money:

**Introduction**

The term plasma, as used in physics, means a stream of ionized particles. This may be compared with a streak of lightning which ionizes the gases of the atmosphere and heats them to incandescence. The electric arc is capable of ionizing both solids and gases. Substances are usu-

Formal definition of plasma (class: stream; differentia: ionized particles), followed by comparison and classification to help au-

ally thought of as existing in three states: liquids, solids, or gases. Ionized substances are sometimes thought of as being in a fourth state of matter.

The plasma torch provides an electric arc between a tungsten electrode and a water-cooled copper nozzle. Gases such as nitrogen, hydrogen, or helium are forced through the arc and nozzle so that they are heated and become ionized; the stream from the nozzle is therefore a plasma stream of ionized particles.

While the plasma torch may be used for both welding and cutting, this report analyzes the cutting properties as related primarily to ferrous metals.

The technical break-through that permitted the cutting of square corners in carbon and low alloy steel occurred several years ago. This was achieved by combining the plasma cutting process with a numerically controlled (N/C) director.

This, in effect, opened the door for industry to apply a new machining concept that minimizes many traditional methods such as sawing, drilling, milling, and shearing. . . .

### Tape Preparation for Plasma Cutting and Related Problems

In N/C flamecutting, several problems are unique to the programming of plasma arc. Most of these problems are caused directly or indirectly by the speed with which plasma cuts or by the bevel generated on the piece part by the plasma arc.

One major problem is the rounding of outside corners and the burning off of sharp end points on plasma cut parts. This is caused by the "following error" of the N/C machine. Following error occurs because the flame cannot keep up with the programmed path of the torch and when a direction change takes place, the flame tends to round the corner. The faster the machine is going, the more pronounced the rounding of corners becomes. In the case of a part with a sharp end point, this can sometimes cause the part to be as much as ⅛ in. (3.175 mm) to ¼ in. (6.35 mm) short of the actual programmed dimension. Following error can be corrected by using programmable pause or dwell codes. With proper use of such codes, it is possible to cut relatively sharp and distinct outside corners.

The cutting of inside peripheral corners and of inside cutouts is another programming problem, also related to the following error. A perfect inside 90° corner is virtually impossible to attain. On one hand, if no pause is programmed in

dience visualize the subject.

Process description to emphasize the function of the device, the plasma torch.

Some background (narrative) on the technical developments that lead to the new machining concept discussed in this report.

Causal analysis of the problems associated with plasma cutting and numerically controlled flamecutting—causes of very observable effects. This causal analysis will lead to a discussion of solutions to these problems.

One solution.

Description of effects

the corner, a rounded or radius effect, sometimes very pronounced, will be produced. On the other hand, if pause is programmed in the corner, often a blow hole or gouge is burned into the part. This not only detracts from appearance but also could cause a point for weakening that could lead to part failures. Experimentation with the individual machine is necessary to determine which method produces the most satisfactory results.

The location of a pierce point and burn-in and burn-out is often the most difficult step of programming any plasma cut part. It is necessary to choose a pierce point for a burn-in to get the most economical use of the raw material. The job is further complicated by the fact that the point that may be the most economical is often not the most feasible from the standpoint of getting a good plasma cut part. For example, burning in on a radius is a bad practice especially on thinner material because often the part drops out before the cut is complete, and a tip is left. The bevel caused by the plasma arc makes it good practice to make the starting and finishing cuts as close to being perpendicular to each other as possible. Semicircle burn-ins can be used on radii and angled lines but they sometimes leave slight gouges in the part.

N/C flamecut machines with kerf compensation capabilities greatly simplify part programming but they also create other problems. Kerf compensation units make it possible for the programmer to program the part exactly as it is shown on the engineering print. There is no need to take into consideration the width of the flame which often varies due to raw material or the burning tip. Kerf compensation is an additional factor which helps multiply the effort of following error on inside corners and cutouts. Kerf compensation is also the major reason for having a burn-out. . . .

N/C tape preparation for these machines is done mainly by computer assist with a minimum amount of manual programming. Our system consists of a minicomputer, 10-cps terminal, 75-cps high speed tape punch and reader, high speed line printer, and a graphic plotter. The system and the necessary software are leased from an N/C computer assist company. A time sharing backup system is also available.

To program a detail part, a text of geometric definitions defining the part from the engineering print and a series of cutting statements are fed into the computer. The computer processes the input and aids in the debugging of the pro-

that can be avoided by an experienced operator.

More causal analysis and description of effects. The audience learns all possible effects of using the tools and gains advice on how to correct some of these effects.

Within the causal analysis, a contrast of the benefits of kerf compensation to the disadvantages of kerf compensation. Since the audience addressed is experienced, it is advised rather than instructed.

A description of the system that the writer recommends is given now that the audience understands what possible effects are beneficial and detrimental.

The writer gives a process description of the system in operation so that the audience can continue

gram. The graphic plotter is used to further debug and verify the program. It is also a valuable aid in determining nesting patterns and economical raw material utilization. After the debugging is complete, machine tapes are pulled and sent to the shop. This process takes anywhere from ½ hour to 3 or 4 hours, depending upon the complexity of the part. Average length of tape is 14 ft (4.2 m). . . .

to weigh the advantages of the recommended system.

### Quality of Finished Parts

An experienced operator will consistently produce completed parts with an edge finish of 200-300 RMS with minimal dross adherence to underside of material. Note: Plasma cut shapes will have an edge angle of from 5° to 7° per inch which is caused by the extreme temperatures generated during the cutting process: 20000° to 60000°.

### Dross Removal

Dross is readily removed with the conventional slagging hammer and/or hand grinders. Slag generally flies off in long sections when shocked by hammer. The cut edge of material tends to be harder than the base material but tests have proven that the H.A.Z. of mild steel-T-1 and Van-80 materials in sizes ¼ in. (6.35 mm) through ¾ in. (19.05 mm) will not exceed 0.040 in. (1.02 mm). . . .

As the report moves toward more specific advice, the writer gives the procedure for dross removal.

### Machine Utilization

Machine utilization is determined by many factors ranging from the skill of the operator through the adequacy and reliability of the material handling system. Torch monitors were installed on each cutting head and machine and torch times (actual firing time per torch) were monitored for a period of 4 months. The results were that machine run time averaged 72% while average torch time was 58%. While this is considered an acceptable level of utilization, we are constantly searching for ways to improve the system.

The results of the writer's tests and the procedure that determined these results. This paragraph again establishes the writer's expertise and thoroughness so that the audience is willing to accept recommendations.

### Conclusion

The plasma method of cutting production parts will result in significant savings in labor and related dollar cost per piece part as contrasted to conventional machining and Oxy-Acetylene cutting.

Care must be taken when committing parts for this process. Only parts that have subse-

The writer's final conclusions and recommendations to the audience. The analysis, descriptions, and procedures given in the report should lead up to this final paragraph. The audience

quent weld operations or those that would be acceptable from a quality standpoint (considering the edge taper) should be selected.

should accept and anticipate the conclusions. Rhetorical patterns are chosen and combined to best support the writer's purpose.

Notice that the writer opens with a definition and a process description of the plasma torch. He then analyzes the problems that this process solves and the problems it may create and interprets his data by identifying causes and speculating about effects. He suggests partial solutions to these problems such as kerf compensation units. Although the process saves money, because several problems remain, the writer suggests that the plasma method be used only when cutting certain parts.

In the analytical report then, we study a problem or process in depth, analyze the characteristics or advantages and disadvantages, and then make a recommendation based on our conclusions and findings.

# The Progress Report

The progress report is usually submitted to a manager or supervisor at a certain point in an on-going project. In progress reports, we review the background of our project for the audience, state the progress we have made so far, and assess where we should go from here. We anticipate any future problems and suggest how to handle them as well as relate problems we might have already faced and overcome. The report helps our audience decide whether to change the direction of our project, allocate further funding or manpower, or stop the project entirely. Although our audience has these options and more, usually we want to appear confident that we are handling the project well. Just as we write a logical and persuasive proposal at the beginning of a project, we write a logical and persuasive progress report usually midway through the project.

For example, in the following progress report, the writer reviews the background of the project, his activities for a certain period of time, and his initial conclusions. Although he may later reassess these early findings, he does offer some tentative conclusions to show that he is going in the right direction toward a solution or recommendation. He also describes what he intends to do before finalizing his conclusions and reassures his audience that he can handle any adjustments or problems that come up (notice that this is the progress report for the project suggested in the proposal in Chapter 12):

MERCEDES MANUFACTURING CORPORATION
POTSDAM, NEW YORK
APRIL 25, 1979

TO:        Thomas P. Clarkson, Vice-President of Manufacturing
FROM:      Martin C. Knox, Director of Machining Division
SUBJECT:   Numerical Control vs. Conventional Machining Progress
           Report

## Background

With increasing sales and constant supply of production space, we must find some mode of operation that will increase production with the existing amount of room. The prime objective of my project is to gather data and evaluate the advantages and disadvantages of numerical control contrasted to conventional machining. I hope that numerical control will be the answer to our problem.

## Activities for the Period

As of now, I have gathered most of the information and am in the process of interpreting it. A major source of data was two companies in central New York, Canastota N/C Corporation and Westlake Manufacturing, both at 119 West Center Street in Westwood City. The managers of these companies were kind enough to let me watch how both types of machines function, and they discussed the advantages of each. They also supplied me with their records of a job they made using conventional and numerically controlled machines. I have come to the following conclusions thus far:

1. A numerically controlled machine is really an electronically designed conventional machine.
2. The personnel required to make a single part for a numerical control machine may be as many as three people while a conventional machine usually needs only one person.
3. The personnel required to run numerical machines need not be experienced, while the operator for conventional machines must have a complete understanding of feeds, speeds, and set-up procedures.
4. Producing parts by numerical control is much faster than conventional machining. The more complex the part, the faster the numerically controlled machining.

## Future Plans

I will make actual cost comparisons between the different kinds of machines from the information gathered from the two companies in central New York. I will do this by using as an example a specialized steam turbine bolt that was produced by both methods. I will cover the descriptions of the parts (blueprints), the quantity produced on each machine, the price per piece related to each method, and the number of man hours worked.

Finally, I will draw conclusions and make recommendation about the practicability of numerical control at Mercedes Manufacturing Corporation.

## Conclusions

The entire project has gone smoothly so far. Despite the time and cost of traveling to another location, I think that the information gained from the trip has been a great asset. Although I supplement this information with that from secondary sources, the tentative conclusions suggested by my findings assure me that we are heading in the right direction. I will complete my report on or before its due date.

A progress report analyzes the success of a project at a point before its completion and makes tentative conclusions about the project. The general tone of a progress report should be one of confidence and competence. Often periodic reports submitted to managers of departments on a monthly, semiannual, or annual basis report "progress" in the form of sales or profits; these are "progress" reports on enterprises that have no definite end.

# The Oral Report

Very often we are asked to present orally the major ideas of a written report, especially if our report was well received or we prepare data for oral delivery only. The conventional format of an oral report is very similar to that of a written report, and we use rhetorical patterns to support that conventional format. However, our delivery and style differ greatly. Also an oral report is perhaps the greatest exercise in audience analysis, since we must be attuned to the needs and interests of our audience before, during, and after we give our presentation.

## Delivery

Since our audience listens to rather than reads our ideas, we must present those ideas clearly. Our audience cannot reread difficult portions of a text; thus we must simplify and repeat key ideas. We must speak slowly enough to be understood. We can determine this rate of speaking by practice beforehand and by watching during the speech to see that the audience is interested and attentive. Normally we can pace our words at about 145 per minute, a pace that varies according to audience and situation.

Although we can deliver an oral report by memorizing it or reading from a manuscript, the best way to present an organized speech and yet maintain eye contact with our audience is to speak from note cards. Note cards allow us to refer to important thoughts and to give exact data or quotations, and also force us to "talk" to the audience. We can be sensitive to the audience's reactions and modify our delivery or repeat our points if the audience looks confused. Eye contact is essential in an oral presentation. Note cards help us be not only well organized but also personal and attentive. Topic or sentence outlines along with any important quotations, statistics, and exact wording are the best content for note cards.

When we deliver an oral report, our physical movements can help us explain ideas and attract our audience's attention. Our gestures should emphasize our ideas. Using physical movement in a purposeful way helps us overcome nervous gestures that detract from our presentation. Physical movements also help us stay attentive and appear confident and enthusiastic about our presentation.

Eliminating our "uhs" and "ums" in our delivery and turning our nervous energy into appropriate gestures such as pointing to a visual aid or turning to various sections of the audience help us appear less nervous than we probably are.

Again, along with a confident and smooth delivery, we must analyze our audience as we prepare for our presentation, be aware of our audience during our presentation and modify our approach should it not be well accepted, and be alert during the question period following the presentation.

## Visual Aids

Visual aids, often much simpler than the graphic aids we might have included in a written report, save time in our presentation when we present data or descriptions. They also clarify our points or help an audience visualize details. They attract the audience's attention and emphasize our most important ideas. However, to use visual aids effectively, we must practice using them to eliminate struggles with a flimsy chart or stuck slide that distract from a presentation. In general, visual aids must be clear and simple, large enough for everyone present to see them, and well coordinated and controlled by us. While they cannot replace a speech, they can support ideas.

**Slides, Film, Videotapes, and Filmstrips.**   If we need realism in our presentation, slides of a location or device let the members of our audience see exactly what we want them to. While films and videotapes interrupt a presentation unless they are short, about 10 minutes, filmstrips and slides can be stopped and started or we can continue to speak over them.

**Blackboards.**   The blackboard, while a common device in the classroom, is only effective in a presentation if we prepare the data on it before our presentation. If we have written at least the major portion of our display ahead of time, we can resist the temptation to turn our back on our audience, and also we can ensure that the data are clear and well organized.

**Flipcharts.**   A flipchart is a large pad of paper attached at the top and propped against an easel. We can prepare charts and graphs ahead of time and turn to each one when appropriate. Even though a flipchart is large, we still must be certain that all our data are visible. We should practice flipping from page to page so that we do not upset the easel during our presentation.

**Handouts.**   Printing material beforehand for each member of the audience ensures that each can see a chart or graph and remember the data on it. The members of the audience feel that they have taken away something tangible from the presentation. However, since when anyone receives a piece of printed matter, the temptation is to read it immediately, we should use handouts only for material that we can distribute after the presentation. In this way, we can keep our audience's attention.

**Transparencies.**  Perhaps the best form of visual aid to display graphs and charts is a transparency projected on a screen by an overhead projector. Transparencies do not distract the audience as do handouts or flipcharts, and we can smoothly and quickly change transparencies. Again, transparencies should not be overly complicated and must be visible from every view in the room, but we can add color or slowly reveal our data for effect. We can turn the projector off while not displaying data and back on at the appropriate time without distracting the audience. We can add overlays to a basic transparency to build ideas or add data.

No matter what visual aids we use, they should be clear and visible, add not distract, supplement not replace our presentation, and be ones we can use with confidence and ease. Practicing use of visual aids is as important as practicing our timing. Finally, any visual aid works better if we prepare the members of the audience for it by telling them the purpose of it.

While a list of suggestions on delivery can never take the place of a speech class or simple practice, most speakers are effective if they remember that the audience is listening, not reading a text. A good oral presentation cannot contain the detail of a written report, but, delivered with enthusiasm and confidence and well organized, can be just as effective.

## Conventional and Created Formats of an Oral Report

**The Introduction.**  In the beginning of an oral report, we must state our topic and our purpose. We must do so in a way that attracts the audience's attention and interest. We should make clear the scope of our report—how much we will say—and our method of organization or plan of delivery—how we will say it. If our audience knows "where" we are going, all listeners will more easily follow. In the introduction, we must establish not only a friendly rapport but also a clear, understandable approach to our material.

For example, in the following portion of an introduction to an oral report given to a group of California businessmen, the speaker personally relates to his audience, attracts attention, and clarifies his subject:

> It's a real treat to be here, not only in the state, but in the very area of the state that originated the most famous proposition [Proposition 13] since Adam said to Eve, "Hey, you're different!"
> . . . Today I want to return the favor by offering you Californians a related proposition, the proposition that we ought to take a hard look at government actions that impose what you might call "nontax taxes." Putting it another way: *"taxes" that aren't really taxes can be worse than taxes that are!*
> Earlier this year, a study of Federal regulatory costs was done for

the Subcommittee on Economic Growth and Stabilization of the Joint Economic Committee of the Congress. This study concluded that the ultimate cost imposed on the private sector by the Federal regulatory agencies is about *twenty times* the direct cost of operating these agencies. . . .

The speaker goes on to quote from sources that explain the effect of "nontaxes."

**The Body.**   In the body of an oral report, we discuss our subject and support our proposition. In an oral report, however, transitions or word "bridges" between our thoughts are essential. Our audience must *hear* when we change direction or leave one thought for another. Often stating how many points we will make and then numbering them as we make them provide this transition. Again we use rhetorical patterns to develop our discussion.

For example, in the body of the speech on "nontaxes," since the speaker has a definite proposal in mind, he uses the rhetorical pattern of procedure. He also numbers the step in the procedure so his audience can "hear" when he begins a new one:

> *First,* a comprehensive economic statement should be required for every significant new legislative or regulatory proposal. Let's never again buy anything unless we know its full cost.
> *Second,* every major new law or regulation should embody what's usually called a "sunset" provision, or a mechanism providing for a phase-out unless the measure is explicitly renewed. . . .
> *Third,* an obligation that should be imposed on government itself is the duty of efficient, skillful, and diligent management of all programs. I'm talking primarily about the "people factor . . .".
> *Fourth,* I think all of us are concerned about different laws and regulations promulgated and enforced by different agencies, that take us in different directions. . . . It would make a lot of sense to have some mechanism to arbitrate when the agencies are independently pushing and pulling in different directions.
> *Finally,* just as government has been holding *our* feet to the fire, we've got to hold *their* feet to the fire. Concerned citizens, and representatives of the business community in particular, have to appear more often, and participate more diligently, in not only the legislative process, but also in budgetary review proceedings. . . .

In an oral report, we have the choices of rhetorical patterns and visual aids that we did in our written report.

**Conclusions and Recommendations.**   As in a written report, at the end of an oral report we summarize or reemphasize our points, draw a logical conclusion, and recommend concrete action. The speaker on "nontaxes" ended his speech as follows:

> We've neglected our golden opportunity to speak up and be heard at appropriations times. Useless, extravagant, counter-productive programs are re-funded through default, through our failure to intervene.
> The tax revolt that erupted here in California is a healthy first step toward better and more efficient government for all Americans. But

let's not use up all our ammunition to cut down direct taxes and government spending. Let's remember the dollars of government spending that are used to impose even larger costs on all of us.

That, in my opinion, is a proposition that we can't afford to turn down!

In an oral report, we use the same conventional and created formats as in a written report, but we must remember that our audience is listening and not reading. We must give special help in understanding and remembering our ideas.

# Summary

Three of the most common types of reports in professional communication are the analytical report, the progress report, and the oral report. Although every report has conventional parts and is organized according to its special audience and purpose, these three types of reports meet essential needs within an organization.

The analytical report describes and analyzes the parts or characteristics of a problem, a situation, a procedure, and such, and makes recommendations about that subject. An analytical report guides our audience through the subject, our approach, and our logical conclusions about the subject.

The progress report, like the proposal, is one step in the communication part of an important project. After our proposal has been accepted, we report our progress at a certain point in the project. We assure our audience that we are handling the project in the best possible way, we offer some tentative conclusions to prepare our audience for our final report, and we describe the steps we will take next.

Often we are asked to present findings orally. Although our delivery of an oral report must be clear and confident, we must organize the report so that the audience can understand and remember our ideas.

Although the three most commonly used formats in any organization are the letter, memo, and report, we cannot depend on these conventional formats for the essential organization of our communications. The organizational or rhetorical patterns we have studied in this text must be combined and adapted to meet the needs and interests of our specific audience and purpose *each time* we "create" a communication.

# Reading
## to Analyze and Discuss

## I

## Safety Evaluation of Renovated Wastewater from a Poultry Processing Plant

*University of Pittsburgh Graduate School of Public Health for the U.S. Environmental Protection Agency*

### Abstract

A three-phase evaluation of reclaimed process wastewater for reuse was undertaken at the Sterling Processing Corporation plant in Oakland, Maryland. The main objective was to evaluate the safety for human consumption of poultry exposed during processing to an average 50 percent mixture of treated well water and reclaimed wastewater. To that end, a determination was made of the ability and reliability of the water reclamation system to deliver satisfactory quality water, and whether the processed poultry would have any excess microbiological or chemical constituents, harmful to human health, as a result of exposure to such water. After the renovation system was optimized (Phase 1), a 2-month study (Phase 2) was instituted, which simulated recycle of renovated water through the poultry plant. Chemical, physical, and microbiological analyses were performed on various water, wastewater, and poultry samples. An experimental chiller, filled with renovated water, was utilized to compare the uptake of such constituents by the processed birds with that resulting from exposure to the chiller in the processing plant using the normally treated well water. An evaluation of the Phase 2 study, as well as other data, leads to the conclusion that the safety of the consumers of the poultry would not be jeopardized if the planned trial period of reuse (Phase 3) were instituted.

### Introduction [in part]

In early 1970 a wastewater reclamation project was initiated at a poultry processing plant in Oakland, Maryland, located in the far western part of that state near the junction of Maryland, West Virginia, and Pennsylvania. An increase in production at the Sterling Processing Company was and is limited by the lack of additional water of acceptable quality. The plant currently slaughters and processes approximately 50,000 birds per day (usually chickens) in an 8-hour operation, utilizing approximately 1,300,000 liters (350,000 gallons) per day of treated well water.

With substantial financial support, initially from the U.S. Department of Interior in January 1971, and later the U.S. Environmental Protection Agency (EPA), a joint project was developed by the Maryland State Department of Health and Mental Hygiene and the Sterling Processing Corporation to design, construct, and study the feasibility of using a wastewater renovation system, utilizing as its raw water source the chlorinated effluent from the second of two aerated lagoons. . . .

**292**

An additional project was proposed by the Maryland State Department of Health and Mental Hygiene and funded by the EPA (Grant No. S803325), the purpose of which was to modify and optimize the reclamation system, to determine the capability and reliability of the system for delivering satisfactory water quality, and to evaluate the exposure of the processed carcasses to constituents that could be harmful to human health. With separate funding (Grant No. R804286), EPA supported the Graduate School of Public Health, University of Pittsburgh, to design, supervise, and perform the sampling and analytical part of the study, as well as evaluate the results from the points of view of both the quality of the renovated water and the processed poultry possibly affected by it. Three phases were planned in this study, the first two of which have been completed. Phase 1 involved the operation of the reclamation plant with a new sand filter, and measurement of those characteristics pertinent to optimizing the process. Phase 2 involved a study of a wide range of physical, chemical and microbiological constituents, both at various points in the reclamation system, as well as in process carcasses chilled with renovated water, but without actual recycle through the plant. . . .

## Discussion [in part]

**Overview.** Poultry, usually chickens, are shipped by truck from the hatchery in Delaware to the Sterling plant and are processed on the day of arrival. Within the plant the birds are slaughtered, scalded, picked, eviscerated, chilled, cut-up and packaged. The poultry processing wastes resulting from the evisceration step are treated by rotary screening for the removal of feathers and viscera. The raw wastewater (sample point A, Fig. 14-1) from the other poultry-processing steps, the refrigeration drains and the plant clean-up passes to a mechanically aerated primary lagoon which is equipped with a grease skimmer. The effluent (sample point L-1-E) from the primary lagoon discharges through a weir trough into a mechanically aerated secondary lagoon (sample point C'), which also contains a grease skimmer. Each lagoon is about 1.8 m deep. The first has a capacity of about 14,000 m$^3$, and the second about 6,000 m$^3$. Their combined retention time normally is 2 to 3 weeks. The effluent from the secondary lagoon is chlorinated as it passes into a combination settling unit and chlorine contact chamber. The chlorinated effluent (sample point C) is discharged through an overflow weir trough to the Little Youghiogheny River. . . .

**Methodology.** The experimental chiller using renovated water was operated on a batch basis. That is, it was filled with renovated water and plant-made ice was added to bring the temperature initially to about 13°C. The 25 carcasses were then lowered into the bath within the drum and rotation begun. Additional ice to maintain 13°C was added. Approximately 15 minutes later more ice was added to reduce the bath temperature to about 1°C, and chilling continued for another 10 minutes.

The carcasses were removed by handling with clean plastic gloves, and both plant and experimental chiller carcasses were treated in the same fashion. They were placed, after draining, in either clean or pre-sterilized (by autoclaving) plastic bags and carried to the laboratory trailer. 1500 ml of either distilled or distilled and sterile water was added to each such bag, which was then shaken for one minute and the water contents poured for analysis. This rinse sampling method has been used primarily for the detection of bacteria in processed poultry, but was employed in this study for both microbiologial and chemical analyses of the carcasses.

Trace metals were analyzed by atomic absorption spectrophotometry using solvent extraction to increase the sensitivity. Reference samples to

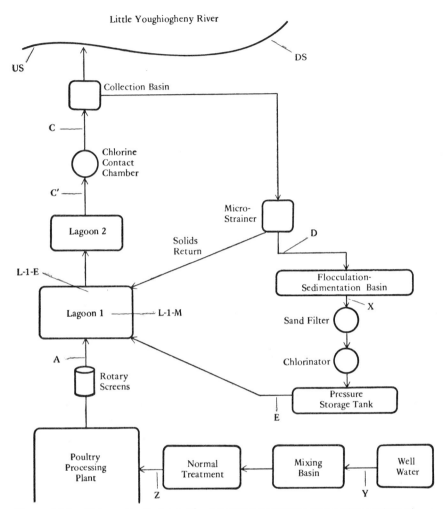

*Figure 14-1.* Schematic diagram of wastewater treatment-renovation system at poultry processing plant with water sampling points. Sample point identification: A, untreated wastewater; L-1-M, lagoon 1, L-1-E, lagoon 1 effluent; C′, lagoon 2 effluent, unchlorinated; C, lagoon 2 effluent, chlorinated; US, river, upstream; DS, river, downstream, D, microstrainer effluent; X, flocculated, settled effluent; E, fully renovated water; Z, normally treated well water. Julian B. Andelman, U.S. Environmental Protection Agency, August 1979.

test and improve, where necessary, the accuracy of the analyses were obtained from the EPA and utilized for trace metals, several other inorganic constituents, and pesticides.

· · ·

NDV is a paramyxovirus that ranges in size from 70 to 120 mm. It is somewhat resistant to adverse environmental conditions, e.g., pH (stable at pH 2 to 12), temperature (stable at ≤ 50°C), light and moisture. Consequently, water can serve as a vehicle for transmission of the virus. NDV is capable of hemagglutinating chicken, guinea pig or human type O red blood cells. In the laboratory it is grown primarily in eggs by inoculation of the allantoic cavity; however, it has been grown also in primary cell (CE) cells. Although NDV has also been grown in continuous primate cell cultures, e.g., HeLa and monkey kidney, a lower titer of virus is produced in the primate cells than in CE cells. The cytopathic effect (CPE) produced by NDV in cell cultures is characterized by syncytium or giant cell formation. Infected cultures can also show hemadsorption when guinea pig or chicken red blood cells are added.

. . .

**Results.** Several specific organic chemicals have been identified, and some quantified in this study. Pesticides were not found in either the renovated or treated well water. Surfactants in the former were well below criterion levels. Several fatty acids were found in the renovated water, but also in the treated well water. In any event, these constitute no human hazard. The maximum concentration of the only halogenated methane found regularly in the renovated water, chloroform, was 3 $\mu$m per liter, well below the approximate median value of 20 found in the EPA National Organic Reconnaissance Survey of U.S. public water supplies. Two phthalates were found in the renovated water, and one of these in the treated well water. As noted previously, both of these, widely used as plasticizers, have been found in potable U.S. municipal water supplies, as well as many natural waters.

. . .

The low bacterial counts in the renovated water, as well as the measurements showing the absence of specific pathogens such as *Salmonella,* demonstrate that bacteria from the water supply do not constitute a risk. The presence of a variety of such bacteria in the plant chiller water or carcasses is certainly not unusual, and they have been shown to build-up rapidly in the water as the poultry contact it. Newcastle Disease Virus that might be expected in the poultry or wastewater could not be found in either. The laboratory die-off experiments with this virus using lagoon water from the Sterling plant indicated that one can expect substantial viral removals in the aerated lagoon system. In view of the approximately 2-week's detention time in the aerated lagoons and the nature of the disinfection processes subsequent to them, which involves two stages of chlorination, this excellent microbiological quality is to be expected. With actual recycle into the plant, this high quality and the additional treatment, including disinfection, would ensure, with a high degree of certainty, that there would be no danger from pathogenic organisms in the reuse of this renovated water. . . .

## Conclusions [in part]

Based on the regularly low or zero bacteriological counts in the renovated water during the Phase 2 study, as well as at other times, the absence of any avian virus that could cause human disease by the enteric route, and the extensive wastewater and water treatment, including four points of chlorination, it is highly unlikely that the contemplated reuse of water at the Sterling Processing Corporation plant would pose a risk of disease from microorganisms to the consumers of the poultry.

The inorganic and physico-chemical characteristics of the renovated water consistently met applicable standards of quality for potable water. A few such parameters, and several non-health related ones, as expected were high, but not at levels that would constitute a threat to human health, even if the water were directly ingested. However, in actual reuse it would receive additional treatment and would not be used as a drinking water supply. A few nonhealth related chemical constituents were at concentrations that could interfere with the optimal operation of the renovations system, and should be adjusted. These include the low pH and high ammonia concentrations. . . .

### Recommendations [in part]

The results of this and previous studies of the wastewater and renovation systems at the Sterling Processing Corporation plant indicate that there are not any apparent concentrations of chemicals or microorganisms that did, or in actual reuse would be likely to build up in the renovated water supply to the point of jeopardizing the health of the consumers of the poultry processed with that water. There are some steps that could and should be taken to minimize and further reduce any possible risks or concerns. It is, therefore, recommended:

1. A trial period of reuse (Phase 3) should be instituted as soon as possible. During this period there should be a full scale monitoring of the renovated water and carcasses as originally planned, followed by a comprehensive evaluation process prior to any permanent reuse.
2. Prior to and/or during the trial period of reuse more extensive organic analyses of the renovated water and other pertinent samples should be performed, with particular focus on chlorinated organics, especially those that might form from the reactions of chlorine with waste products from the poultry. Assessments should then be made of the possible health significance of any identified and quantified organics, including a determination of the likely impact of the quantities to which the consumers of the poultry would be exposed. . . .

# *Questions for Discussion*

1. What rhetorical patterns are used to develop the abstract? Does the abstract indicate the problem, the methodology, and the solution and recommendations?

2. What rhetorical patterns are used to develop the introduction? Does the introduction serve as a "bridge" into the report?

3. What rhetorical patterns are used to develop the portions of the discussion section we have here? How well are they combined to analyze and solve the problem?

4. What rhetorical patterns are included in the conclusions and recommendations?

5. What are the audience and purpose of this report? Are the style, structure (conventional and created), and content appropriate to this audience and purpose?

# *Exercises and Assignments*

**1. Problem:** Your audience is pleased with the memo report you wrote for exercise 3 in the last chapter. He or she asks you to do a further investigation of the problem, perhaps analyzing how the problem is handled on other campuses or assessing how the problem may affect other areas of the college. The audience asks you to investigate further and write an analytical report on the subject.

Assignment:

Broaden your approach to the subject and write the analytical report your audience has requested.

**2. Problem:** Your local college has two campuses: one in the center of town and one 2 miles away. The central campus houses the administration and the school of arts and sciences. The other campus houses the schools of business and engineering. You work for a consulting firm that has been hired by the president of the college to investigate the possibility of adding a busline between the two campuses. At the moment, most students drive between campuses, and parking is a problem. You are in charge of investigating the problem and recommending a solution. The busline may or may not be the best answer.

*Comment:* Use your own campus for inspiration and your observations and imagination to supplement the details of this project. You probably should consider cost, traffic flow, possible stop locations, alternative parking areas, and so on.

Assignment:

1. Midway through your investigation, write a progress report to your supervisor reporting: the information you have gathered so far, your methods, any problems, your adherence to schedule, your conclusions if any so far, and so on. Use both the conventional report format and a created format appropriate to your audience and purpose.
2. Write the report to the president of the college. Use both the conventional report format and a created format appropriate to your audience and purpose.
3. Prepare a brief (10-minute) oral report on your findings to present to the Board of Trustees of the college. You might change your approach slightly for this new audience.

**3. Problem:** Consider how a project that you are working on this semester in one of your technical, business, or science classes might be presented in a report to a hypothetical business or technical audience.

Assignment:

Using the research you have done this semester in another class, write a proposal, a progress report, and a final report to a hypothetical audience. Try to be as persuasive as possible and end your final report with concrete recommendations. Use the conventional report formats and the rhetorical patterns appropriate to your audience and purpose. Be sure to include graphic aids where necessary.

# Credits

*(continued from p. iv)*

p. 17, Reading ID, "Handwriting on the Sky," *The Nation*, Vol. 226, No. 6 (February 11, 1978), p. 131. Copyright 1978 The Nation Associates. Used with permission.

p. 29, extract, Jacob Bronowski, *A Sense of the Future* (Cambridge: MIT Press), p. 255.

pp. 37–9, Reading I, J.B.S. Haldane, from pp. 20–24 in *Possible Worlds* (New York: Harper & Row, 1928). Copyright © 1928 by Harper & Row, Publishers, Inc.; renewed 1956 by J.B.S. Haldane. Reprinted by permission of the publisher, the Author's Literary Estate, and Chatto & Windus.

p. 46, extract, Sigmund Freud, "An Introduction to Psychoanalysis," in *A General Introduction to Psychoanalysis*. Excerpted from *The Autobiography of Science*, edited by Forest Ray Moulton and Justus J. Schifferes. Copyright © 1945, 1960 by Justus J. Schifferes. Reprinted by permission of Doubleday & Company, Inc.

p. 50, extract, *Guide for the Care and Use of Laboratory Animals*, Committee on Care and Use of Laboratory Animals of the Institute of Laboratory Animal Resources, National Research Council (Washington, D.C.: U.S. Department of Health, Education, and Welfare, 1978), p. 3

p. 52, extract, John Read, "Chemistry," *What is Science? Twelve Eminent Scientists and Philosophers Explain Their Various Fields to the Layman*, James R. Newman, Ed. (New York: Simon & Schuster, 1955), pp. 154–155.

pp. 52–3, extract, Isaac Asimov, *Adding a Dimension: Seventeen Essays on the History of Science* (New York: Doubleday, 1964), pp. 85–86.

p. 53, extract, Khalil Taraman and Dale Valvo, "Glass Cutting by Ultrasonic Grinding," Technical Paper, Society of Manufacturing Engineers, 1975, p. 1.

p. 54, extract, *IMC Methylamines: A Complete Guide*. Courtesy of International Minerals and Chemical Corporation, 1979, p. 6.

p. 56, Reading IA, *Policy Implications of Medical Information Systems* (Washington, D.C.: Office of Technology Assessment, November 1977), p. 3.

p. 56, Reading IB, LaRoux K. Gillespie, "Machinability as Related to Precision Miniature Parts," Technical Report, Society of Manufacturing Engineers, 1976, pp. 1–2.

pp. 56–7, Reading IC, Mitch Waldrop, "Science Update," *Chemical and Engineering News*, Vol. 56, No. 23 (June 5, 1978), pp. 20–21.

p. 61, extract, René Laënnec, "On Mediate Auscultation," 1819. Excerpted from *The Autobiography of Science*, edited by Forest Ray Moulton and Justus J. Schifferes. Copyright © 1945, 1960 by Justus J. Schifferes. Reprinted by permission of Doubleday & Company, Inc.

p. 62, extract, John T. Shea, "Structure and Design Considerations of the Vanguard SLV-5 Magnetic Field Satellite," Technical Note D-707 (NASA: Washington, D.C. 1961), p. 2.

pp. 66–67, extract, "Bell System Reports Chicago Lightwave Communications Evaluation a Success," News from the Bell System, public relations release, May 1978. Work done by Bell Labs.

pp. 67-8, extract, "Seaway Pipeline Company, Seaway Freeport Dock Tanker Unloading System," report by Phillips Petroleum Company, Bartlesville, Okla., 1979, pp. 1–2.

p. 69 extract, "Flexible-Pile Thermal Sealant," Lyndon B. Johnson Space Center, Houston, Texas, *NASA Tech Briefs*, Vol. 1, No. 3 (Fall 1976), p. 405.

p. 74, Reading I, David A. Sonstegard, Larry S. Matthews, and Herbert Kaufer, "The Surgical Replacement of the Human Knee Joint," *Scientific American*, Vol. 238, No. 1 (January 1978), pp. 44–45. Copyright © 1977 by *Scientific American*, Inc. All rights reserved.

pp. 75–6, Reading II, L. R. Ertle and W. H. Day, "USDA Quarantine Facility, Newark, Delaware," *Facilities for Insect Research and Production*, Norman C. Leppla and Tom R. Ashley, Eds., U.S. Department of Agriculture, Technical Bulletin 1576, June 1978, pp. 49–51.

p. 84, extract, I. P. Pavlov, "General Types of Animal and Human Higher Nervous Activity," *Selected Works* (Moscow: Foreign Languages Publishing House), pp. 318–319.

pp. 85–6, extract, Loren Eiseley, "The Golden Alphabet," in *The Unexpected Universe* (New York: Harcourt, Brace & World, 1969), pp. 121–122.

p. 87, extract, Carl Sagan, *The Dragons of Eden: Speculations on the Evolution of Human Intelligence* (New York: Ballatine Books, 1977), pp. 166–167.

pp. 88–9, extract, Dames & Moore, "Sources of Energy," *Engineering Bulletin 49*, pp. 24–26. Permission granted by Dames & Moore, publisher.

pp. 89–90, extract, Isaac Asimov, *Extraterrestrial Civilizations* (New York: Crown Publishers, 1979), pp. 157–158. Copyright © 1979 by Isaac Asimov. Used by permission of Crown Publishers, Inc.

p. 91, extract, Dan B. Walker, "Plants in the Hostile Atmosphere," *Natural History*, Vol. 87, No. 6 (June-July 1978), p. 76, 79. With permission from *Natural History*, June/July 1978. Copyright The American Museum of Natural History, 1978.

p. 93, extract, courtesy Syska & Hennessy.

p. 96, Figure 5-1, Brian K. Lambert and R. M. Sundaram, "Comparative Performance of Cemented Carbide and Titanium Carbide Coated Cutting Tools," Technical Paper, Society of Manufacturing Engineers, 1976. Copyright © 1976 Society of Manufacturing Engineers. All rights reserved.

p. 98, Reading I, David H. Hubel, "The Brain," *Scientific American*, Vol. 241, No. 3 (September 1979), p. 46.

pp. 100-1, Figures 5-2 and 5-3, Faye Serio, photographer.

p. 104, extract, Lamarck, "New Theories of Evolution," in *Zoological Philosophy*, trans. Hugh S. Elliot (London: Macmillan and Co. Ltd., 1809).

p. 105, extract, Thomas Malthus, "An Essay on the Principle of Population, as It Affects the Future Improvement of Society," 1978.

pp. 105–6, extract, "Couples Want Fewer Children," in "Science and the Citizen," *Scientific American*, Vol. 241, No. 4 (October 1979), p. 72.

p. 106, extract, "Barriers to Urban Economic Development," U.S. Government Report, May 1978, p. viii.

pp. 106–7, extract, L. A. McReynolds, *Small Engines—An Energy Perspective*. Reprinted with permission, © 1978 Society of Automotive Engineers, Inc.

p. 107, extract, *The Future of Productivity* (Washington, D.C.: National Center for Productivity and Quality of Working Life, Winter 1977), p. 132.

pp. 108–9, extract, David W. Fraser and Joseph E. McDade, "Legionellosis,' *Scientific American*, Vol. 241, No. 4 (October 1979), pp. 85–86.

pp. 109–10, extract, Timothy L. Montgomery and David J. Rose, "Some Institutional Problems of the U.S. Nuclear Industry," *Technology Review*, Vol. 81, No. 5 (March/April 1979), pp. 56–58.

pp. 110–11, extract, Orrin Riley *et al.*, "Effect of Air Pollution Regulations on Highway Construction and Maintenance," National Cooperative Highway Research Program Report 191, 1978, p. 20.

p. 111, extract, Marvin Harris, *Cannibals and Kings: The Origins of Culture* (New York: Vintage, 1977), pp. 53–54.

p. 113, Table 6-1, Syska & Hennessy, Inc., Engineers, "Brownouts and Voltage Reductions," Technical Letter, Vol. 24, No. 9 (June 1974), 2.

p. 113, Table 6-2, Frank M. Butrick, "Engineering and Trouble Shooting of Spade Drill Applications," Technical Paper, Society of Manufacturing Engineers, 1976, p. 12. Copyright © 1976 Society of Manufacturing Engineers. All rights reserved.

p. 115, Reading I, T. E. Davidson & Associate, Inc., personal correspondence.

pp. 115–16, Reading II, Lewis Thomas, *The Medusa and the Snail: More Notes of a Biology Watcher* (New York: Viking, 1979), pp. 19–24. Copyright © 1974–1979 by Lewis Thomas. Originally appeared in *The New England Journal of Medicine.* Adapted by permission of Viking Penguin Inc.

p. 117, Figure 6-1, Faye Serio, photographer.

pp. 121–2, extract, Charles Darwin, "The Origin of Species," 1895, in John Warren Knedler, Jr., Ed., *Masterworks of Science*, Vol. 3 (New York: McGraw-Hill, 1973), pp. 69–70.

p. 127, extract, David G. Lygre, *Life Manipulation: From Test-Tube Babies to Aging* (New York: Walker, 1979), pp. 50–51.

pp. 129–30, extract, Los Alamos Technical Review LASL 79-16, Los Alamos Scientific Laboratory, Los Alamos, New Mexico, p. 1.

pp. 130–1, extract, *Common File System Primer*, Los Alamos LA 7499M, Vol. 3, Los Alamos Scientific Laboratory, Los Alamos, New Mexico, July 1979, p. 5-1.

pp. 131–2 and Table 7-1, Sandra A. Mamrak and Paul D. Amer, *Computer Science & Technology: A Methodology for the Selection of Interactive Computer Services* (Washington, D.C.: U.S. Department of Commerce, January 1979), pp. 4–5, and Table 7-1

p. 134, Reading I, *The Business of Filmmaking*, Eastman Kodak Company, 1978, p. 3. Reproduced with the permission of Eastman Kodak Company.

pp. 134–5, Reading II, Isaac Asimov, "He's Not My Type," copyright © 1963 by Mercury Press, Inc. Reprinted from The Magazine of Fantasy and Science Fiction, which appears in the book *Adding a Dimension* by Isaac Asimov. Reprinted by permission of Doubleday & Company, Inc.

pp. 137, 138, Figures 7-1 and 7-2, from *How Does It Work?* by Richard M. Koff and illustrations by Richard E. Rooman. Copyright © 1961 by Richard M. Koff. Reproduced by permission of Doubleday & Company, Inc.

pp. 144–5, extract, Maurice Maeterlinck, *The Life of the Bee* (New York: Dodd, Mead & Company, 1901).

p. 148, extract, "Analysis of Trends in Policy and Technology," *Power Engineering*, Vol. 83, No. 5 (May 1979), 16.

pp. 148–9, extract, "What Went Wrong with the Three Mile Island Reacter?" *Physics Today*, Vol. 32, No. 6 (June 1979), pp. 77–78. By permission of the American Institute of Physics.

pp. 188, 189, extracts, Operating and Troubleshooting Instructions, RF-310 SSB Transceiver, Instruction Manual. Courtesy of Harris Corporation, RF Division.

pp. 190, 191, Warnings, cautions, and notes all taken from Fisher Controls manuals and instruction sheets.

p. 192, Figure 10-1, Reprinted with the permission of Fisher Controls Company. Figure 10-2, Courtesy Harris Corporation, RF Division.

p. 193, Table 10-1, Courtesy Harris Corporation, RF Division.

p. 194, Reading I, Receiver Protector Test. Courtesy of Harris Corporation, RF Division.

p. 195, Reading II, "Fundamentals of Roscoe," Clarkson Computing Center. Courtesy Applied Data Research, Inc.

pp. 201, 202–3, 203–4, extracts, Procedures adapted from insurance memos and personal correspondence.

p. 202, extract, Marvin W. Burnham, "True Tool Path Considerations," Technical Paper, Society of Manufacturing Engineers, 1976, p. 7.

pp. 206–8, 210, extracts, Construction Specifications, Special Provisions, Department of Transportation, Sacramento, California.

pp. 209–10, *Report Writing Guide.* Courtesy of Lester B. Knight & Associates, Inc.

p. 212, Reading I, adapted from student experience and *FM22-5 Field Manual Drill and Ceremonies* (Washington, D.C.: Department of the Army, November 12, 1971).

p. 213, Reading II, from Special Provisions for State Highway Construction, May 1978, September 1977, February 1978.

p. 226, extract, "Syllogism as Reason" example courtesy of Phillips Petroleum Company.

pp. 228–9, extract, Adversary enthymeme example courtesy of Lester B. Knight & Associates, Inc.

p. 230, Courtesy of Phillips Petroleum Company.

p. 231, extract, Robert D. Reckert, "Reflections on Professional Development in Iowa," DeWild Grant Reckert & Associates.

p. 232, extract, Patricia Curtis, "New Debate Over Experimenting with Animals," *New York Times Magazine,* December 31, 1978.

p. 238, Reading I, Courtesy of Phillips Petroleum Company.

pp. 239–40, Reading II, Courtesy of Richardson, Runden & Company, Inc.

p. 241, Figure 12-1, Faye Serio, photographer.

p. 256, extract, Drainage System memo, adapted from memo courtesy Department of Public Works, Kansas City, Missouri.

p. 262, Table 13-2, courtesy of Fisher Controls Company.

p. 277, Reading I, Courtesy Department of Public Works, Kansas City, Missouri.

pp. 281–5, extract, John E. Barger, "Plasma Arc Cutting," Technical Paper MR76-712. Copyright © 1976 by the Society of Manufacturing Engineers, Deerborn, Michigan.

pp. 289–91, extracts, Richard F. Schubert, Vice Chairman, Bethlehem Steel Corporation, "Taxes That Aren't Really Taxes Can Be Worse Than Taxes That Are!" October 1978. Courtesy Bethlehem Steel Corporation.

pp. 292–6, Reading I and Figure 14-1, Julian B. Andelman, "Safety Evaluation of Renovated Wastewater from a Poultry Processing Plant," (Cincinnati, Ohio: U.S. Environmental Protection Agency, August 1979).

# Index